A Drinker's Guide to Pure Water

A Drinker's Guide to Pure Water

◆

Is Your Water Safe

Zalman P. Saperstein

iUniverse, Inc.
New York Lincoln Shanghai

A Drinker's Guide to Pure Water
Is Your Water Safe

iUniverse books may be ordered through booksellers or by contacting:

iUniverse
2021 Pine Lake Road, Suite 100
Lincoln, NE 68512
www.iuniverse.com
1-800-Authors (1-800-288-4677)

ISBN-13: 978-0-595-39518-7 (pbk)
ISBN-13: 978-0-595-83917-9 (ebk)
ISBN-10: 0-595-39518-X (pbk)
ISBN-10: 0-595-83917-7 (ebk)

Printed in the United States of America

This book is for all living things, which deserve to flourish in an unpolluted world that nature and our creator intended. Our children and our children's children deserve to inherit a world that is less contaminated than it is now. We must all help to bring this goal about.

Contents

Acknowledgements

Special thanks go to Sheila Sabrey-Saperstein, PhD, wife and loving soul mate, who patiently edited and suggested many improvements. John Zeile, a retired Lutheran minister and a wonderful man, also contributed extensively by reading the drafts, editing, and helping to reduce the rancor that sometimes surfaced. Thanks also to Kurt Pagel, a retired forest ranger and close friend, for his specific ideas and input.

Thanks to the tireless and dedicated advocates, organizations, individuals, scientists, and workers who constantly strive to protect our environment. Their ongoing contributions and investigations provided the backbone for this book. Their work must continue so that we can attain a better world for present and future inhabitants.

Thanks also to iUniverse, and particularly Janet Noddings for her guidance and counsel during the publishing process, and David Bernardi for his thorough and helpful editing.

Finally, we are grateful to the few elected politicians who are dedicated to protecting our environment and health. We often rely upon them to promote necessary support for environmental legislation. Perhaps the future will see more politicians become dedicated to the environmental causes that benefit all of us.

Author's Comment

Much of the information used in this book was derived from extensive research conducted by the author, and the sources have been referenced throughout. However, this book is for the users of drinking water, and that means everybody. I deliberately avoided page after page of footnotes, citing sources, and instead included a reference where important sources are listed. Additional sources are included in the text as appropriate. The intent of this was to make the book more readable.

We all are responsible for our environment and the quality of our water. This requires a basic understanding of how we endanger our water quality, and what we must do to prevent additional toxic contamination. Environmental scientists and scientific specialists cannot carry this responsibility alone. We must all be involved.

Please tolerate my passion and occasional rancor about certain actions by corporations, government officials, and we, average citizens, as we seem to often ignore the consequences of our actions. I am as guilty as anyone. Unless we are aware of the actions that contaminate our drinking water we will continue to inflict harm upon ourselves and others, albeit often unconsciously. The passion and rancor that surfaces reflects the desire to stimulate our constructive actions. Apathy and complacency will doom us. Passion is required. Constructive action is imperative.

Introduction and Overview

The purpose of this book is to bring to the reader's attention to important facts and issues about our drinking water, stimulating concern and interest among all of us affected by government and corporate actions that are generally unknown. The subject of this book, drinking water, deserves consistent and prominent attention in our news media. Such attention by the media, as well as by governmental agencies and elected politicians, is usually lacking. When the media gives attention it is often in obscure publications, in seldom-read sections of newspapers, on non-commercial cable or satellite television shows, or on public radio broadcasts. News concerning water contamination should always be on the "front page," because it affects all of us. The news media must inform the public and do the investigative work necessary to publicize otherwise hidden news about our water quality and the pervasive contamination that exists.

Chapter 1 is an overview of the last seventy-five years, the time when water contamination became the critical problem that it now is. Chapter 2 deals with the actions that have taken place to severely contaminate Earth's waters. A brief historical summary in Chapter 3 reviews some important events and describes the legacy of environmental pioneers to whom we are indebted, as well as some of the polluters who were, and are, responsible for life-threatening water contamination. Chapter 4 delves into some of the major causes of water contamination. Chapter 5 summarizes recent studies and investigations that substantiate that drinking water contaminants cause serious health consequences. Chapter 6 poses the question, "Can government protect our environment and water quality?" The answer is laden with bureaucratic incompetence, procrastination, self-serving politics, and corporate indifference. Chapter 7 presents a number of recent events from different regions of our country that demonstrate how widespread drinking water contamination is and the crucial need to stay informed to prevent contamination. Chapter 8 recommends some actions we can take to protect ourselves from water contamination, including the formation of "Water Action Groups." Finally, Chapter 9 asks how and what we can change to foster environmental awareness and responsible action to protect our water. One unconventional approach to achieve this objective is "Mission Earth."

Water is something that we all are intimately familiar with, something based on our direct physical experiences. It is the most abundant natural substance on earth. We feel its wetness as we splash in an ocean or lake, or as we bathe and shower. We have all appreciated the thirst-quenching gift of water as it soothes a parched mouth. A cool piece of ice stimulates our tongues and lips as nothing else can. We glory in a spring rain that feeds the budding flowers in our garden. Our eyes shed water when we experience intense joy or grief. Vaporous water emanating from a boiling kettle whistles a merry tune as we prepare for a festive party or an afternoon cup of tea. Water is, fortunately, readily available to most of us, and we usually take its presence for granted, as we do the air that we breathe. Water and air are very special gifts from nature that must be treasured and not taken as a given. Water binds our planet, creating a commonality shared by all living organisms and vegetation. As is air, water too is indispensable.

From the time life originated on Earth until a very short while ago, the water molecule, which human, animal, and plant life depends upon, consisted of a single hydrogen atom and two oxygen atoms combined into a molecule and containing whatever chemical elements or compounds that Nature included. The oceans are and have been saline for eons, and that is fortunate for ocean fish and sea life that thrive in saline, aqueous ocean habitats. Water in its natural form was all that was present on earth. That has changed.

At the onset of the Bronze Age, around 4,400 BC, and into the Iron Age, about 500 BC, the products that we made and used were from natural substances that were available in or upon the earth, from animals, plants, and trees. As we progressed into the Industrial Revolution about 350 years ago, we learned how to refine certain minerals and create metallic alloys, such as steel, and products that served more diversified purposes, including agricultural machines, cookware, and weapons. As the natural and physical sciences evolved during the 1700s, 1800s, and early 1900s, the things we produced were still from natural substances, but they required more processing.

The advent of the "Chemical Age" began, escalating around 1930, merely two-and-a-half generations ago. PCBs (polychlorinated biphenyls) and the insecticide DDT (dichloro diphenyl trichlor ethane) were two of the first widely used manmade chemicals, both developed in the 1930s. Since then, several hundreds of thousands of manmade chemical substances were developed and are in common use in many consumer, industrial, and military products. *These manmade chemical substances never existed until created by the efforts of budding inventors and industrial corporations during the last seventy-five years.* This represents the beginning of what I term the "Manmade Toxic Contamination Age." This is now.

The Manmade Toxic Contamination Age is the current time of widespread toxic and carcinogenic contamination of the pure molecular combination of hydrogen and oxygen, known as water. The three atoms are no longer alone in a molecule of fresh water. True, fresh water does not exist naturally, except in ancient polar ice regions or glaciers, in some subterranean sources, or perhaps in some very remote area on earth. Pure water is obtained by processing ordinary contaminated water in special research laboratories, or in facilities where absolute purity is essential, such as for medical and pharmaceutical products. The molecules of "natural" water now on earth contain myriad manmade toxic contaminants, which also reside in our bodies, endangering our health and the well being of infants yet to be born.

1

The Drinking Water Crisis

Drinking water is something we easily take for granted. We unconsciously assume that if our water is clear as it comes out of a tap, it is safe to drink. Circumstances occasionally shock us into the realization that this is a dangerous assumption. Warnings given by local authorities to boil water before drinking it to kill the bacteria that somehow contaminated our water represent such an occasion. This is uncommon. We generally drink, cook, bathe, and tend our gardens and plants with water without a second thought about purity. That certainly is the way it should be. Purity should be the norm. Unfortunately, that is not the reality.

Water, the basic elixir for all life, has also become a health hazard, as it flows from our household taps into our water glasses, cooking utensils, baths, showers, and, most importantly, our bodies! This Jekyll and Hyde duplicity has resulted from the gradual contamination and poisoning of our water supplies with toxic and carcinogenic chemical compounds. This has resulted from human activities, mostly originating during the last seventy-five years, since the approximate beginning of the Manmade Toxic Contamination Age.

Prior to 1930, the year when highly-toxic PCBs were first used industrially, most toxic substances contained in water originated from naturally occurring sources, such as minerals, metallic substances, and fossil fuels in our earth. The refining of petroleum products, mining operations, the burning of coal and wood for heat and electric power generation, industrial operations, and waste disposal generated some contamination, but the volume was substantially less than now, because population and consumption levels were much lower. Automobiles were comparatively few in number, and public transportation was much more common. Even Los Angeles had electric powered streetcars and buses!

Most of the durable goods in the 1930s were cars, washing machines, and "ice boxes" (not refrigerators, because mechanical refrigeration was just becoming commercial).

Farm implements used primary metals (iron and copper) and natural woods. Home construction materials came from natural substances only. However, some natural pigments (including lead oxide) found in many paints contaminated homes and buildings and caused serious illness and disabilities, particularly damage to the central nervous system and brain. Plastic products were in their infancy, but they are now part of almost everything we use, from food containers to automobiles.

Herbicides, insecticides, and pesticides produced from synthetic chemical compounds were uncommon. Toxic natural elements, such as arsenic, were contained in pesticides. Fertilizers and chemical nutrients were not typically used for agriculture or home gardens, and when used they were derived from natural, not manmade, materials (crops that enhance nitrogen fixation using crop rotation, etc.). Agricultural operations were mostly organic, because the manmade chemical compounds used later did not yet exist.

The cosmetics and home-products corporations were relatively small operations with limited offerings that used mostly natural substances, plants, and animal products. Ice, not synthetic refrigerants later used in mechanical refrigeration machines, kept our foods cool.

Cotton, wool, silk fabrics, and other natural materials were the only materials used for clothing, furniture, and rugs. Synthetic polyester fabrics were not available for common use. Rubber was a natural extract from trees and not synthetically derived. Synthetic fluorine and chlorine containing polymers were virtually unknown. Food wrappings and containers were plain paper or naturally waxed papers, not plastics, as they commonly are today.

Home medications and drugs were limited in kind and usually consisted of simple blends of naturally occurring substances. Aspirin was such a product, originally derived from the bark of a specific evergreen tree. We now use over-the-counter and prescription drugs made from synthetic chemical compounds that we consume and then discharge, via our bodies' waste-removal systems, into sewage or landfills. Trace amounts of these chemical compounds return via ground water to water that we ultimately use and drink.

We all unconsciously expect that the toxic trace remains from materials and substances that we consume and use are eliminated by the sewage treatment or natural decomposition process. They are not. The sewage effluent streams that discharge into rivers, lakes, and oceans contain residual bacterial and chemical substances that re-contaminate our water supplies and land. Trillions of gallons of treated toxic water are generated daily by municipal or private sewerage treatment plants throughout the U.S. alone. Some of the contaminated effluents discharged into nearby rivers, lakes, or oceans tend to concentrate and accumulate and others biodegrade. Semi-solid sewage sludge materials, are composed of body wastes and toxic chemicals, are sewage treatment products and are incinerated, and then stored as ash, in open landfills, or more commonly, stored as raw sludge often close to populated areas. The raw sludge or ash is laden with contaminants that can eventually leach into land and into our water supplies including reservoirs and aquifers. Today's sewerage-treatment technology does not adequately protect the purity of our water supplies, even in our advanced country. This cycle of contamination from sewage waste to potable water represents one of the hidden dangers to our health as populations increase, and high-density development mushrooms. The generally obscure relationship between discharge from sewerage-treatment plants, and consumable water supplies is not generally recognized by the public. This relationship will be discussed in more detail later.

After 1930, the development of hundreds of thousands of manmade chemical compounds ensued, and the exponential growth of new industries and products followed. The natural chemical elements, listed in the periodic table of any high school chemistry text, have been and are now being used in combinations and blends that never existed on Earth before the early 1930s. Just look around at the things that you touch, use, eat, and smell, and ask yourself which of these contain chemical substances that were created since your grandparents were born. The answer is "virtually everything."

Indeed, approximately 400,000 manmade chemical compounds now exist, and at least 90% of these were developed and first used after 1930. Many thousands are in common use today. Those of you who are younger than seventy-five have had the dubious opportunity of being the first humans in Earth's history to keep company with many of these "alien visitors" to our planet. Is it logical to accept these "alien" chemical compounds to be compatible with life that has evolved over millions of years in their absence?

Certainly, some of the "alien visitors" are welcome. They have helped cure diseases, prevented starvation, and provided employment opportunities for many millions of people. Others have provided improved quality of life, improved life for disabled individuals, improved survival rates among cancer patients, and extended life for the terminally ill. Medical advances have reduced death from infectious diseases and have extended our life expectancy. We are sometimes able to survive otherwise deadly cancers or diseases because of new medical treatments. We all are grateful for the benefits that have resulted from human endeavors. However, we must ask ourselves how many of the debilitating diseases of human and animal life are the result of environmental and water contamination.

Are we on an endless treadmill, trying to cure or ameliorate diseases that we are causing because of environmental contamination? Substantial evidence supports this conclusion, as the reader of this book will realize.

One pertinent observation pertains to childhood cancers. According to medical history records, childhood cancers were rare before 1940 (records were not systematically kept until 1930) and have been on the increase since. *According to the 2002 U.S. Mortality Public Use Data from the National Institute of Health, childhood cancers are now the second-leading cause of death (accidents being first) of children aged one to fourteen years.* Cancer accounted for 11.6% of the deaths of one-to fourteen-year-old children in the U.S. in 2000. This single fact must raise our suspicions that something relatively new has happened to cause this terrible situation. The Manmade Toxic Chemical Contamination Age is responsible, and the increasing incidence of childhood cancer is very strong evidence, if not proof. The unborn and young children are the innocent victims.

In the general population, cancer deaths nearly exceed those from heart disease. *The infant mortality rate in the U.S. (6.43 per 1,000 live births) is higher than in forty-one other countries.* (Singapore is the lowest, at 2.29.) *Even the average life span in the US (77.85 years) is lower than in forty-seven other countries.* (Andorra is the highest, at 83.5 years.)

Our country consumes more energy and manufactured products than any other. Our consumption creates more waste materials than any other country, and therein is the reason for our relatively high infant mortality rate and comparatively low life expectancy. Increased waste creates more water contamination, air pollution, and land pollution.

The waste we create, whether inadvertently or deliberately, creates a continuous stream of toxic and carcinogenic substances. *More than 50% of the so-called managed waste is not biodegradable and is stored in landfills,* where it steadily accumulates and drains contaminants into our water supplies. Our material wealth as the world's leading industrial power is also creating a legacy of contamination. Since the beginning of the 1930s, our waste has become more toxic as the use of new manmade chemical substances has proliferated.

We must recognize the adverse effects that the vast array of recent chemical compounds and associated products cause as they contaminate air, water, land, food, vegetation, and animals. They are becoming integral contaminants in all living things, including all human fetuses as they develop in their mothers' wombs. We owe it to the unborn children to question what effect these contaminants will have on their health and mortality, as well as on the future of all life on earth. As we reap the benefits of technology, we must also become aware of potential pitfalls. It is unwise and dangerous to assume that all new things are beneficial, useful, and harmless. They clearly are not.

Professor Alan E. Flanigan, my deeply cherished mentor as an undergrad and graduate student at UCLA, taught his students many gems of wisdom. One of the most valuable is expressed in these few words: "One test is worth a thousand expert opinions." This precautionary advice means that the result of a *single experiment* is capable of revealing whether or not your explanation, hypothesis, or fledgling theory is potentially valid. This simple but powerful admonition is unfortunately often ignored by corporations, governments, pseudo-scientists, and us, the public, as we plunge evermore into a morass of chemical contamination about which we know very little (and generally ignore). We are assured by words that small amounts of this, that, or the other chemical or drug is safe, only to learn later that the experts were wrong. The critical experiments are often never performed; instead, companies solicit the opinions of "experts," who proclaim the safety of a product that actually causes direct harm to individuals or contaminates our environment, damaging large populations of animals and people. Today's environment is much too complex and interdependent to allow hastily designed and ill-conceived commercial products to contaminate our landscape and water. We must require experimentation to validate the safety of new chemical substances and products before they are considered for production. Scientists who are independent from profit driven forces alone must have this responsibility, and they must be completely independent from corporate and political influence. We must always err on the side of public health and the protection of life.

We have all become the unwitting subjects of a multitude of ongoing corporate and government "experiments." The *toxic chemical substances* contained within or generated by the products we use are the independent variables, and *our health* is the dependent variable. We are analogous to laboratory rats, subjects of a massive experiment that has been proceeding in an escalating manner for seventy-five years. Many products have and are being developed, manufactured, and sold are done so any regard for the potential long-term harm they will inflict. We use and dispose of various products often without any concern about environmental impact. We generally do not even think about the effect that the disposal of waste has on the land and the purity of our drinking water supplies. Everything that we use becomes, or generates, waste (often a direct byproduct, such as the exhaust from an automobile, boat, or lawn mower engine). All waste is contamination that enters our environment, and potentially our drinking water. All contamination has the potential to cause disease and death among living things—plants, animals, and humans.

PCBs are but one lingering example. The closely related PBDE's (polybromi-
nated diphenyl esters) and DuPont's Teflon (and the chemicals used to produce
it) are the latest toxic health threats that have been found in drinking water and
human body fluids. A recent investigation conducted by John Hopkins medical
researchers revealed that *toxic PFOA (used to produce Teflon) was in the umbilical-
cord blood of 298 out of 300 newborns tested.* Medical researchers collected umbili-
cal-cord blood samples over a five-month period beginning in late 2004, and are
now investigating the potential health consequences in the infants whose blood
contains PFOA. John Hopkins Medical School researchers should analyze umbil-
ical-cord blood for the hundreds, or thousands, of other synthetic chemicals that
contaminate our environment. What would they find?

Hundreds, and probably thousands, of known or suspected disease-causing
contaminants are potentially in our drinking water. Yet, most of us seldom give
this a thought. Our tax-paid government bureaucrats are even more delinquent
and generally take years to identify and regulate harmful substances. The Environmental Protection Agency (EPA), the Centers for Disease Control and Prevention (CDC), the National Institute of Health (NIH), and the sometimes-
involved Food and Drug Administration (FDA) are all *slow-acting government
bureaucracies*, usually administered by untrained political appointees. The
appointees are often individuals who were previously engaged in corporate activities. Is there any wonder that many regulatory actions designed to limit contamination are often politically delayed for years, rejected as economically unfeasible,
or even eliminated from regulatory consideration?

Today, there are literally no constraints on what corporations can manufacture and sell for general public use. There are no governmental agencies with the
responsibility and authority to regulate new substances and products before corporations release them for general public use. The FDA regulates new drugs and
determines the standards for drug safety and food purity, but these regulations do
not apply to the majority of new things sold to an unsuspecting public, which
may cause contamination and harm to our health. How can the regulatory agencies of the federal and state governments constantly fail to restrict the use of an
ever-increasing number of toxic and carcinogenic substances? Why do these agencies allow the use of substances that inevitably contaminate our drinking water
with toxic chemicals?

Human diseases and the premature death of loved ones remind us of the result
of contaminated drinking water. Cancer and disabling diseases in newborn and
very young children suggest that something is awry, and medical evidence con-
firms this. Stillbirths are another indicator. Toxins and carcinogens that we may

ingest with water, food, and air are causing ongoing harm to human and all life. Virtually all of these contaminating toxins and carcinogens are synthetic chemical compounds in products and substances that we use and consume. *They have been contaminating our environment, and us, for a miniscule fraction of man's history.* It is impossible to know the ultimate damage that will occur to living organisms and human life because of this exposure. We are in the early stage of this potentially irreversible experiment, and the ultimate outcome is yet to unfold.

If it is not already too late, what are we going to do about it? The outcome, which may be determined in the next seventy-five years or less, will depend upon our actions. We must not stand by and let the political policy makers, corporations, and our own apathy continue to allow our earth to become more contaminated. We must begin to reverse the trends of the last seventy-five years if we expect to preserve and protect all life.

2

Water Is Our Most Valuable Resource

Our Planet's Constant Water Supply

The world's water supply, like fossil oil, is not limitless. For all practical purposes, water is a zero sum game. This means that any water on our planet is here to stay as a solid (ice), gas (steam), or liquid, adsorbed or absorbed into other materials. Our planet's water can neither be created nor destroyed (except by a nuclear reaction or other reactions involving nascent hydrogen) only changed (in what scientists term "phase," i.e., to liquid, solid, or gas) or combined with or released from other substances.

All water on Earth and its atmosphere has been residing here and in our atmosphere for billions, of years. However, the distribution of water and the impurities that it contains constantly change. Therefore, as contaminants enter Earth's water sources, the purity of the water decreases. Certain contaminants remain indefinitely and others may decompose or disintegrate at varying rates into harmless constituents. The Earth's water will never increase in purity as long as we persist in our polluting ways. Natural pollution also occurs and adds to the contamination that we create, and certain natural chemical elements are dangerous carcinogens and toxins, such as arsenic, lead, mercury, uranium, and other elemental substances that we must also take care to avoid.

Water Facts and Usage

Nearly everything that we consume contains some water. Without water, life as we know it would be impossible. Most animals, including humans, are about 90% water by weight. Most vegetation is at least 80% water by weight. Some sea animals are more than 95% water. Our bodies contain over 50,000 *miles* of capillaries, veins, and arteries, which transport blood (more than 50% of which is water, which is constantly being renewed) to every tissue and organ. Without this self-regulating plumbing system and its watery contents, we would all become dry fossils or a collection of dust, bones, and debris.

In the U.S. alone, we use an average of 60 gallons of water per day, per capita, creating a personal demand of about 300 million gallons per day in the state of Wisconsin alone. The average 2 quarts of water intake that we consume in our food and water drink per day is essential to sustain life. The daily total use of household water in the U.S. exceeds 90,000 trillion gallons. This does not include industrial and energy use, which is many times more. The total estimated use in the U.S. is between 300 and 500 trillion gallons per day. The world use undoubtedly exceeds 1,000 trillion gallons per day, and the demand increases as the population of Earth increases. The use is staggering, yet in some parts of the world, there is little no available potable water. Perhaps 20% of the world's water is for *essential* applications, particularly in our country. The remainder is needlessly used for unnecessary purposes. Think about the many ways that we waste water. Compile a daily list and attempt to reduce your waste.

The United Nations has estimated that at least 3 billion people in the world suffer because of insufficient and polluted water, which is often deadly or a source of chronic disease. Pregnant women who have access only to certain types of contaminated water have increased likelihood of miscarriages and stillbirths and a high incidence of physical and neurological defects in their surviving offspring. These often cause premature death. They and their families can become victims of water contamination without being aware of the cause. Young children are especially susceptible, since they are at a developmental stage of life and small in body mass. Lower concentrations of harmful substances are able to induce greater damage than in adults.

Throughout the U.S., we depend upon water from municipal water supplies and private wells. Some may be unfit for human consumption. Municipal water departments supply water in accordance with EPA and FDA standards, but they too may be limited by constraints in budget, technologically (meaning they are unable to detect and control contaminant levels in the effluent streams), and water quality standards. You must remember that EPA, FDA, and state regulatory standards are the outgrowth of bureaucratic decisions, not necessarily derived from scientific facts and data.

The maximum concentration limits (MCL), for example, *specified for 119 regulated (measured and controlled) chemicals in municipal drinking water*, are generally based upon testing limitations, dictated by economic or other considerations rather than the best scientific judgment and knowledge available. Furthermore, the regulated chemical substances are but a small fraction of all the substances known or suspected to contaminate our drinking water and cause health problems. Furthermore, the health effects of different combinations of toxic contaminants are essentially unknown.

The MCLs for individual contaminants are contrived as if they alone were present. Yet they never are, as you can verify by reading any water quality report from any municipality or water utility in our country. (A copy of an actual water quality report is in the Appendix. Did you know your drinking water could contain the toxic chemicals listed?) The complex simultaneous interactions of many contaminants in our bodies (from the water we drink) are almost totally ignored by regulatory agencies responsible for protecting our water quality. Very little research has been done to identify the most dangerous interactions, and even less has been done to impede them.

Small municipal water departments and sanitary districts are simply unable to test for the myriad harmful substances, even if they wanted to, because of technological or budgetary limitations. Owners of private wells generally do not have their water purity tested except for hardness, iron, and harmful bacteria. The cost of a comprehensive chemical analysis would be prohibitive for most of us. The majority of known chemical toxins that are present in ground water are not subject to any analysis for their presence in our drinking water. Our foods also will become contaminated by the water that feeds them, and the water used to process them. Our foods and water are seldom if ever tested for the presence of toxic trace impurities like pesticides, herbicides, and fungicides, which will lead to biological damage in our bodies. Reporting of "unregulated" chemical contaminants by water treatment plants is not mandatory. And as for regulated chemicals, the interactions of multiple, toxic trace-chemicals (from unregulated chemicals in our bodies) are ignored, or neglected, by regulatory agencies throughout the U.S. and most nations in the world.

Environmental contamination determines our health, because it ultimately governs the chemical composition of our blood, organs, bones, and tissues—in short, every part of our bodies. The old phrase "You are what you eat" is more correctly stated "You are what your water (blood) chemistry is." The air, earth, and water on our planet determine the contaminants that enter our bodies. Nothing is more critical to the survival of all life forms than our water, and all life forms must have uncontaminated water, suitable for the sustenance of life.

3

Some Historic Background

Early Environmental Activists

Rachael Carson's classic book *Silent Spring*, published in 1962, portrayed the devastating destruction of life caused by the highly toxic herbicide DDT (a PCB/dioxin-type synthetic chemical, also similar to "Agent Orange," used by our government during the Vietnam War to defoliate the jungles of Vietnam). Her book was the first significant public warning about the lurking dangers of contamination from synthetic chemical compounds. Her warnings, given about forty-five years ago, have gone largely unheeded by multinational corporations and governments, as new chemical compounds proliferate and untested consumer products ultimately find their way into our water and foods. The carcinogenic gasoline additive, MTBE, which was used as a replacement for toxic lead compounds, is a recent example of the rush to commercialize a product before possible adverse health consequences are fully evaluated. Many communities in our country are now discovering carcinogenic MTBE in their water supplies from accidental spillage, leaking gasoline storage tanks, or vehicle exhaust emissions. The rush to commercialize this product with little or no concern for health consequences is the type of irresponsible corporate/government behavior that must cease. Unfortunately, examples that illustrate this type of corporate and government neglect are all too common.

In large measure, because of Rachel Carson's dedicated work on behalf of all living things, environmental awareness began to blossom. Prior to Carson's work, Aldo Leopold, a dedicated naturalist and environmental educator from the University of Wisconsin, had fostered an appreciation of the beauty and benefits of nature. He is known as the "Father of Wildlife Management." His treasured book, *A Sand County Almanac*, published one year following his death in 1948, is one of the most inspiring and beautiful books about nature ever written. Both *Silent Spring* and *A Sand County Almanac* are required reading for all of us who are serious about understanding and protecting our natural environmental heritage.

Another Wisconsin environmental pioneer followed the path that Aldo Leopold and Rachel Carson blazed, but in a different direction. He was our country's preeminent "environmental politician," Wisconsin born and raised Gaylord Nelson. He understood the importance of environmental sustenance (as had both Leopold and Carson) and he used his political abilities to educate the people of Wisconsin and our nation on the subject. In 1969, as a United States senator, Gaylord Nelson originated the concept of Earth Day, and his actions on behalf of our environment led to the adoption of Earth Day as an annual event. His political leadership demonstrates what a dedicated "people's politician" can accomplish to the benefit all of our citizens. No individual politician has ever done as much good for our environment as Senator Gaylord Nelson. Where are the Senator Nelsons of today among the politicians that we elect to represent us? Former Vice President Al Gore stands out as the most preeminent politician who understands and advocates for environmental sustainability. We must listen and support what he is telling us. We must elect politicians who champion and support the preservation of Earth's precious resources and who will resolutely lead us into a new beginning of environmental awareness and preservation.

Individual Citizen Action

As Senator Nelson was working at a high echelon of government for environmental protection, a high school graduate and working mother was working to uncover a deadly water contamination disaster in her California community. This community was the unknowing collective victim of corporate deception, and the woman's name was Erin Brockovich. Her courageous individual efforts during the 1970s helped the victims of massive ground-water pollution caused by a toxic and carcinogenic chemical that induced suspected birth defects, infant deaths, and chronic diseases. A movie depicting these events (*Erin Brockovich*, starring Julia Roberts) later brought the crisis to the world's attention. An energy production corporation, Pacific Gas and Electric (PG&E), used a form of chromium (hexavalent) in its water-cooling system, even though the carcinogenic and toxic properties had been published and were recognized by scientific health professionals. The contaminated cooling water directly mixed with ground water and contaminated the aquifer that provided drinking water from wells. PG&E persisted in using the toxin for about twenty years, causing extensive water contamination and disease.

Thanks to Erin Brockovich's individual activity as an advocate for the affected people to ferret out the truth and publicize the cause of serious health problems among the poor people in her community, which she worked at tirelessly and initially without pay, the people eventually prevailed, and the affected families received a $333 million settlement. Yet, death and permanent debilitating disease is never recoverable, and people continued to become sick years later from the toxic water that they drank, bathed in, and used for everyday purposes.

Facing additional claims of health problems caused in three California counties, the PG&E corporation declared bankruptcy in 2001 in an attempt to avoid additional financial losses. Finally, in early 2006, PG&E settled with about 1,100 victims of their corporate contamination and paid a $295 million settlement. (PG&E thereby avoided a Los Angeles Superior Court trial, after a federal court refused to transfer the case to federal jurisdiction where the financial damage award could have been substantially lower.) After several decades of opposition to all claims against them for their toxic contamination, this energy corporation relented, but it has not admitted responsibility for health problems, and it had paid a total of over $600 million to innocent victims and their survivors.

Beyond Our Borders

One of the worst chemical disasters in history resulted from the neglect of another major U.S. corporation at a chemical plant in Bhopal, India, in 1984. The people of the city of Bhopal have yet to recover from the chemical poisoning inflicted when an unsafe chemical processing plant released a highly toxic cloud of chemical poisons over a vast area. The results: severe and permanent environmental damage, over 6,000 deaths, and many thousands who have endured and are still suffering from chronic and terminal cancers, respiratory, and neurological diseases. Ground water supplies are still contaminated and will remain so for many years. The multinational corporation responsible, Union Carbide, has not fully compensated those who suffered even after more than two decades. Nor can they ever compensate those who have yet to be born with physical and mental disorders or others who are doomed to premature death. The CEO of Union Carbide, who authorized the design and use of an unsafe plant design and was criminally indicted for so doing, has avoided prosecution by refusing to return to India and face trial. The average compensation paid to the thousands of innocent victims was about $550 each. Legal actions are still proceeding at the time of this writing. Victims in Bhopal, India, are still demonstrating, demanding safe drinking water twenty-two years later to replace the toxin-contaminated water they are still forced to consume.

The PCB Saga

Another example of corporate indifference and negligence is very familiar to the people in Door County and the Fox River communities of North Eastern Wisconsin, including Green Bay and Appleton. The highly toxic and carcinogenic chemical substances known as PCBs (polychlorinated biphenyls) are all too familiar to Wisconsinites and millions of others throughout the U.S. who are still plagued by severe problems related to these chemicals. The history of PCBs is a meaningful example of how corporations and government collaborate to avoid responsibility for toxic contamination. For this reason, several pages are devoted to this example. Our understanding may help us prevent future similar contaminations.

The contamination caused by PCBs in our lakes, rivers, and water supplies have wrought death and disease upon industrial workers as well as an unsuspecting populace for over seventy-five years, ever since PCBs were developed, commercialized, and applied in a variety of industries. Animal and human lives have been lost, and countless men, women, and children have ingested PCBs in water or food (particularly lake and river fish), becoming seriously ill or even dying because of the exposure.

The dangers remain because of the stable nature of the substance, its widespread use, and government/industrial laxity in taking remedial actions. Chemical Industry Archive data (available online from the Environmental Working Group) reports that many billions of pounds of PCBs have been discharged or dumped by industrial users into streams, rivers, lakes, land, and air in various regions of the U.S., as well as in many other regions in the world, over a period of about fifty-five years.

The Wisconsin Department of Natural Resources estimates that 500–900 thousand pounds of PCBs remain in the sediment of bodies of water in northeastern Wisconsin (and that it migrates in the waters of Green Bay, Lake Michigan, and the Fox River). Corporate ignorance or indifference was and is the main reason for this environmental disaster, and governmental shielding of the guilty corporations has allowed the contamination to persist and spread, severely threatening the health of hundreds of thousands of people and creatures throughout the area many decades after the toxic nature of PCBs became well known. This neglect is absolutely sinful and criminal.

In 1996, the U.S. Environmental Protection Agency issued a report entitled, The Public Health Implications of PCB Exposures. The report states the following:

> Human health studies are discussed in this paper that indicate: (1) reproductive function may be disrupted by exposures to PCBs; (2) neurobehavioral and developmental deficits occur in newborns and continue through school age children who had in-utero exposure to PCBs; (3) other systemic effects, e.g., self-reported liver disease and diabetes, and immune system risks may be associated with elevated serum levels of PCBs; and (4) increased cancer risks are associated with PCB exposures.

Ten years later, the health dangers described in the U.S. EPA report persist due to PCBs that continue to contaminate water, soils, and food (particularly fish). PCBs accumulate in our bodies as we unsuspectingly ingest them. How can the EPA apparently forget their warning and allow PCB to continue poison drinking and recreational waters in many regions of our country?

One of the world's largest corporations (General Electric Corporation) was an early user of PCBs, and the corporate executives became aware of the toxic and disease-causing nature of PCBs long before the public became aware of it. GE used PCBs in liquid coolants and lubricants for power generating equipment from the early 1930s until the early 1970s. The GE executives involved with this industrial use became aware of the toxic and biologically hazardous nature of the synthetic chemical during the 1930s, when many employees became ill from exposure and many sustained liver and skin diseases. GE initially failed to report the observations to public-health officials, and many more years passed before GE and others stopped using the substance.

GE eventually acknowledged PCBs as hazardous materials after consistent denials. By 1972, the disposal of manufacturing equipment containing PCBs required special permits. The disposal became government regulated and costly to accomplish.

Before 1929, PCB did not even exist. The Swann Chemical Company developed PCB in 1929, after which the GE Corp. became a major user. In 1935, the Monsanto Industrial Chemical Company acquired the Swann Chemical Company and became the patent owner of PCBs. They continued to produce PCBs and sell them worldwide until 1978. Even though the Monsanto Corporate executives responsible for the business and technical decisions knew (for nearly thirty years before production was banned and ceased in the U.S.) that the synthetic chemical compound was likely to cause disease in animals and humans, they continued its production. Initially, the lack of knowledge concerning the potential health affects of PCBs could justify continued production and sales. But, by the mid 1950s Monsanto's numerous internal memoranda revealed that their executives were fully aware of the toxic and disease-causing effects of PCBs, but they chose to continue production and sales nonetheless.

Even the U.S. Navy rejected the use of a Monsanto product containing PCB in 1956 after their testing revealed that it was too toxic to use. Nonetheless, production and sales continued to worldwide industrial users without abatement. (The Navy's decision to reject the product was not made public at the time.) Monsanto occasionally provided precautionary advice regarding potential health problems to the commercial users of PCBs starting in the 1950s. Most customers simply ignored their corporation's "inherent" responsibility to assure that they were using a safe substance and blindly applied the product as they decided was appropriate. [Refer to the Chemical Industry Archives listed in the Appendix for complete information on Monsanto's and other PCB involvements from about 1935 on.]

However, in 1966, independent research by Swiss scientists revealed that they had found PCBs in human hair, fish, bird eggs, and pine needles. Regulations against the use of PCBs in Switzerland resulted soon thereafter. Monsanto executives disputed and attempted to discredit this independent scientific work. However, another independent scientific study in 1968 reported that certain birds and fish contained PCBs. The birds and fish had habitats along the California Coast and in the Pacific Ocean at several different locations, including Puget Sound and Baja, California. The University of California at Berkley scientist who did the investigative work published his results in the scientific magazine *Nature* (vol. 220, 14 December 1968). Dr. Robert Riseborough, the research scientist, wrote that chlorinated biphenyls (PCBs) have three distinct negative impacts:

1. It is widely spread by air and water and therefore an uncontrollable pollutant.

2. It is a toxic substance with no permissible allowable levels. This could be seen in the case of the Peregrine falcon. It caused the extinction of that creature by inducing hepatatic (liver) enzymes, leading to reproductive weakness (thin-walled, fragile shells).

3. It is a toxic substance that endangers humankind itself, the implication being that the Peregrine falcon is a leading indicator of things to come.

Monsanto executives continued to publicly reject all negative findings and ignored the public-health issues. Production and commercial use of PCBs still flourished.

A Monsanto report written in 1969 stated that about 80 million pounds of PCBs (with varying degrees of chlorination) were produced annually at that time. By 1967, an estimated 2 trillion pounds had been manufactured by Monsanto, and it was probable that over 90% had entered the environment, particularly by lake, river, and ground water. Altogether, it is likely that over 5 trillion pounds of PCBs and related substances were produced, sold, and distributed by Monsanto all over the world, from 1935 until about 1980. The stability of PCBs creates an everlasting presence of this toxic substance, and the damage to living things is incalculable. By the year 1969, internal Monsanto memos proved that the corporate executives were preparing for the curtailment of production and probable clean up action at manufacturing locations. Meanwhile, production and sales continued.

In August of 1970, Monsanto learned from the Alabama Department of Public Health (AWIC) that fish taken from a nearby creek (where one of their two PCB manufacturing operations were located) had fifty-five times the legal limit of PCBs set by AWIC. The Food and Drug Administration (FDA), which conducted the investigation, alerted Alabama, which in turn told Monsanto. The FDA initially withheld their findings from the public, even though the local waterways were commercially and recreationally fished. Finally, after a three-month delay, the public was notified about the fish and water contamination caused by PCBs, by an article published in a local newspaper by Alabama Public Health officials—not by the FDA. The news reports were, however, given a favorable media twist orchestrated by Monsanto and Alabama. The reports suggested that there was no reason for the public to be concerned. Fishing for the contaminated fish went on as usual in the PCB-laden Alabama waterways. Production and sales of PCBs continued.

The Monsanto corporate records concerning PCBs represent a repository of apparent executive deception and disregard for public health by an irresponsibly managed corporation over a period of several decades. Over two hundred variations of PCB-like chemical compounds (most in liquid form) were invented and commercialized from about 1930 to the early 1970s, many (by Monsanto and others) even after the toxic and disease causation was recognized and disclosed in reputable medical and biological research journals. These PCB-like chemicals include DDT, DES, dioxin, and other chlorine-containing synthetic chemicals. By the early 1970s, many chlorinated chemical compounds similar to PCBs, such as dioxin, DES, and DDT, became associated with (by corporate insiders and independent medical researchers) serious health problems, including various forms of liver disease, brain abnormalities, arrested physical and mental development, skin maladies, and cancers.

After lab animal tests revealed PCBs were carcinogenic, the U.S. EPA classified them as "probable carcinogens." DES (a synthetic hormone) used by pregnant women and for farm animals (as a supplement for enhancing cattle growth or milk production) was ultimately banned by the U.S. Food and Drug Administration in 1971, but PCBs and other similar substances were unaffected by the FDA order.

Industry continued to develop, manufacture, and sell many similar chemicals, including PCBs/dioxin, for a variety of applications until 1977, when the manufacture and use of PCBs was restricted (but not banned) by the U.S. EPA. Many of the paper manufacturers in Wisconsin used and discharged PCB-contaminated water knowingly into the Fox River for several years after the substance was widely suspected to cause disease. PCBs were employed to manufacture carbonless copy paper and other paper products. These practices continued into the early 1980s, nearly twenty years after the deadly effects of PCB had been made public by scientific studies and at least forty years after Monsanto and GE internally knew about its disease causing potential.

In 1979, Congress enacted a law to cease all production of PCBs in the U.S., but not the use of existing stocks. This occurred approximately fourteen years after the Swiss and University of California studies had been public. By 1979, extensive water contamination and human disease and death, including spontaneous miscarriages and cancers, had occurred in the unsuspecting public and in workers who were exposed or ingested sufficient quantities of PCB. Birds and fish populations were and are sickened and killed by PCBs, as are their natural predators. Industrial exposure and the human consumption of fish, waterfowl, and drinking water supplies continue to endanger the public and the residual PCBs (which are very slow to decompose and accumulate in animals including humans) are longlasting threats to health and life. The continued contamination of our water and soil exemplifies corporate greed, government indifference, and bureaucratic incompetence that unfortunately are all too common.

Unanswered Questions

Are you ready for some tough questions?

• Why did Monsanto allow production of PCBs when their corporate officers were aware of the dangers to public health and animal and plant life?

• Why did the paper companies in Wisconsin use PCBs in the manufacture of products for more than twenty-five years, beginning in 1954, when they knew that the wastewater discharges would contaminate the water and the fish within it, causing serious health problems in adults, children, fetuses, and animals that ate the contaminated fish or consumed the toxic water?

• Why did the use of existing stocks of PCBs continue even after production of PCBs *was banned* in the U.S. in 1979?

• Why did the political administrations in Wisconsin of Thompson, McCallum, and Doyle, and the elected assemblymen and senators over *the last thirty years*, allow PCB contamination to persist since the production ban?

• Why did paper companies continue to use the deadly chemical, even after production *was banned?*

• Why does PCB contamination remain in northeastern Wisconsin after *more than seventeen years has lapsed* since our elected Wisconsin government officials in the Thompson administration and paper industry executives agreed to a comprehensive PCB removal program?

• Why have the Federal Government, EPA, and FDA allowed many of the PCB hotspots and polluters throughout our country to avoid their responsibilities to the citizens of our country not to harm us?

The PCB saga has now been unfolding for nearly seventy-five years. As the toxic master of ceremonies to the Manmade Toxic Contamination Age, this manmade chemical continues to cause illness, and death, in humans and animals. Disease and death caused by the PCBs carried in waters and foods are as certain as disease and death caused by the ravages of a hurricane or tsunami, whose waves of water create a naturally induced calamity which always grabs the headlines and people's attention. The manmade, poisonous, and carcinogenic chemical contaminants in our water and foods seldom grab our attention. They are the unseen causes of disease that are the products of corporate and government negligence and our own apathy or lack of knowledge.

A substantial number of pages in this book are devoted to the issue of PCB contamination. This was a conscious choice, because the PCB saga demonstrates that if corporations and government had acted on a known problem quickly, they could have prevented environmental and personal tragedies that occurred and are still occurring. Unfortunately, selfish corporations, and apparently acquiescent governments, fostered and allowed the ongoing production and contamination caused by this carcinogenic toxin for many decades after its toxicity and harm was established and well documented. The long-lasting poisons still are contaminating waters and are afflicting people and animal life in our country and elsewhere in the world.

The important lesson taught by this ongoing environmental tragedy is this: *Neither governments nor corporations can be trusted to protect the public from toxic contamination. Public oversight and meaningful regulations must be established by independent environmental organizations, not by politically appointed and corporate influenced-bureaucrats, or even our elected representatives in government.* Until independent oversight really is established, we will continue to experience similar tragedies, as newer, even more toxic chemical substances are developed and inflicted upon an unsuspecting public.

The Superfund Isn't Super

"The Superfund" is a special environmental cleanup fund that the Federal government established in the 1970s to pay for the cleanup of dangerously contaminated areas in our country. The current Bush administration and Congress have under-funded the Superfund, to the point where it cannot accomplish the decontamination work needed at hundreds of sites in our country. Politicians have allowed corporations to escape their financial obligations to fund required cleanup actions. Government agencies have twiddled their thumbs and are allowing unsuspecting children, pregnant mothers, unborn children, and the rest of us to suffer from life-threatening diseases caused by these toxic sites. We have seen irreversible environmental and water quality damage to lakes, rivers, water supplies, and food. Unknown numbers of lives have been exposed to avoidable disease and death caused by PCB pollution. Why does our federal government knowingly allow such deadly contamination to exist in our waters, foods, and air? Wake up America, and tell our elected representatives, "No more!"

Government officials are often incapable or lack the political will to respond to the needs of the people. Hurricane Katrina recently disabled the New Orleans water and sewage treatment systems, causing untold ground water pollution that will persist for many years. Toxic chemicals, household products, oil refinery discharges, waste debris, fertilizers, vehicle fluids, and pharmaceutical drugs have saturated the soils and polluted the ground water. What will we learn from this disaster? The people of New Orleans and our country must speak out for our own protection. Pure water is our most precious long-term need, and only the people can force the politicians and our governmental bodies to do what is required to assure our pure water supply.

Many politicians and legislators decry legal actions that arise when the public becomes aware of environmental health problems and individuals attempt to recover damages. Why is it that these same politicians seldom object to the pollution that many corporations unleash upon an unsuspecting public? Could it be that these politicians have their campaign treasure troves supplied by corporate special interests and lobbyists, who expect favors in return? One favor is to continue to contaminate as usual. Profits benefit.

It is imperative that we, the people, become informed about the presence and causes of water contamination and take necessary actions to protect ourselves, and our loved ones, from harm caused by contaminated water. We must eliminate the complicity of governments and corporations, which continue to ignore the urgency of the problems related to water purity. We must shake loose from our apathy and ignorance of water contamination problems and demand proper corrective response from our elected government officials.

4

How Water Becomes Deadly: What You Don't Know Can Cause Harm

A Glimpse Back

Recall the fact that water is neither created nor destroyed and exists as a finite and fixed amount on Earth. It does, however, change in form, distribution, ease of availability, and purity. At some time in the history of our planet, all the water was as pure and uncontaminated as was possible from natural processes. Imagine for a moment that you lived 2,000 years ago in what is now Door County, Wisconsin, or some other once pristine region that has since been spoiled by excessive developments. You were walking along the shoreline looking for fish for your next meal. While walking, you became thirsty and instinctively crouched on the shore and started drinking the pure azure liquid. You reached into the water and plucked a small fish out that would become your next meal. The abundance of pure fresh water and uncontaminated fish provided a bounty, nature's gifts to human life.

Water Today

Fast forward to the present. Would you dare drink water directly out of the shoreline water anywhere in Door County as your possible distant ancestor did 2,000 years ago? Would you have done seventy-five years ago? Fifty years ago? Sheila, my wife, did as a child while visiting from Chicago. The water is still there, but now it is generally contaminated and unfit to drink. If we drink water directly from Green Bay or Lake Michigan, we could now become infected with bacteria, and would very likely ingest a host of toxic, and carcinogenic manmade synthetic chemicals that pollute the water (albeit in trace or even micro-trace concentrations that are detectable by special instruments).

Any fish we catch today in Green Bay or Lake Michigan may contain PCBs in its tissues, as well as other toxic chemicals. Warning signs advise us against the consumption of fish caught in these contaminated waters, and even the EPA often warns against consumption. We instinctively know that drinking shoreline water in Door County, or from most lakes, rivers, and streams virtually everywhere in the U.S. is unsafe. We may have read the warnings about bacteria that have caused numerous beach closings in Door County and elsewhere. Other toxic contaminants such as PCB, VOCs (volatile organic compounds), MTBE (a recent gasoline additive), phosphates (mainly from fertilizers), nitrates (mainly from fertilizers and sewage waste), mercury compounds (mainly from coal-burning power plants and chlorine production), herbicides, insecticides, and petroleum contaminants now pollute the waters and sediments everywhere and can cause a host of diseases and debilitating maladies.

Literally thousands of manmade chemical compounds and materials are carried into lakes and streams via storm waters or surface waters that have become polluted as the waters have flowed over contaminated land, sidewalks, roads, parking lots, and structures. Industrial pollutants have entered our water supply the same way. Disease-causing bacteria are discharged or seep into lakes and rivers from numerous municipal sewage-treatment plants and individual septic systems. Also discharged are traces of innumerable chemical substances that we dispose of by washing them down a sink or flushing them down a toilet, deliberately dumping them, or unknowingly leaking onto land (from cars and machines). These traces eventually drain into septic systems or municipal sewage treatment plants or are flushed and carried by storm waters on a new journey to streams, rivers, and lakes, where additional water and land pollution ensues.

Airborne pollutants such as those emanating from power plants, automobiles, cigarettes, aircraft exhaust, and boat-motor exhaust are transferred to the ground and surface waters by gravitational or wind forces and by rain or snow. The contaminants in air, rain, snowflakes, lakes, rivers, and streams become an integral part of our land and water ecosystem. Some of the contaminants return to subterranean aquifers and reservoirs, adding to the existing contamination. Even water treatment plants cannot completely purify our tap water, as any water-quality report shows. The technology used today is incapable of removing all toxic and carcinogenic contaminants. Treated sewage water that discharges into lakes and rivers contains biological contaminants and manmade substances unaffected by the treatment process because of the technological limitations of the equipment and processes employed. Sewage sludge removed to landfills also contains residual toxic contaminants that leach into ground water and aquifers. The effluent streams from sewage treatment plants contaminate the water they flow into rendering recreational and fishing activities unsafe. Contaminated surface waters that recharge aquifers re-contaminate the aquifers with the chemical and biological contaminants that are not removable from effluents or sewage sludge. Significant improvements in sewage and water treatment technologies are necessary to minimize this source of repetitive re-contamination.

Evaporation of surface waters occurs and the subsequent condensation as rain or snow returns water to the surface. Contaminants that are present in the environment become components of "new" air, rain, and snow that are again transferred into aquifers, reservoirs, rivers, lakes, streams and the earth as the cycle continues. This water cycle constantly degrades our aquifers and open water supplies as long as new contaminants are available. Unfortunately, new or renewed contaminant sources are readily available because we humans continue to create and use new materials, many of which are toxic and biologically dangerous to the sustenance of life. These substances are constantly entering the finite water supplies on Earth and thereby endangering human and all other forms of life. As we humans extract more water from the aquifers, the remaining water becomes more concentrated with contaminants.

Concentrations of toxic and bio-hazardous substances in Earth's water are immutably increasing. This is precisely why endangered species are an important signal for ecologists and biologists, who understand that the gradual elimination of one species is a potential threat to all life, including humans. The oil problem of today will be child's play compared to the day of reckoning when the Earth's water supplies become too hazardous for human consumption. If you are a skeptic, as many will be, think about our attitudes about oil and gasoline just a few decades ago. I remember buying gas for 20 cents a gallon for my first car in 1950. No problem, huh? Think again.

Our water purity was sometimes worrisome then, but generally it was not a serious problem even during the 1940s and 1950s. Times have changed, and so has the quantity and quality of water that we consume. We must not take our water supply and quality for granted, and we must become proactive to protect it for present and future generations. To do so, we must first understand the ways that water becomes contaminated and our roles in these processes. We must then take responsibility to prevent contamination.

How We Contribute to Contamination

Consider the pathway of a drop of water *before* it exits our kitchen or yard faucet. Where did that drop originate, and what intervening events might have influenced its purity as it flowed into a glass or container of water that we drank or used for some other purpose? How did this liquid enter our bodies and reach every cell that depended upon it for survival? What happens when that water drop contains substances that, although they are present in trace or micro-trace amounts, are inconsistent with the life and normal functioning of the interdependent cells that form our bodies?

Most of the water used in the United States comes through pipelines to users from various sources of fresh water aquifers, lakes, or reservoirs. Wisconsin, for example, derives most of its tap water from aquifers and Lake Michigan. Door County, in Wisconsin, is totally dependent upon water from aquifers, which are particularly vulnerable to ground-water contamination because of the highly permeable fractured rocky surface and subsurface soils. The underground aquifers are analogous to large, porous earthen reservoirs, lakes, or rivers of water that originally formed inside Earth eons ago. They are "recharged" by the natural water cycle if the ambient conditions and rate of recharge (rain and snow) and discharge (use) allows. As an aquifer recharges, the water within it tends to become progressively less pure—because of manmade substances that accumulate in the water that recharges it and because of the residual contaminants remaining in the aquifer. The build-up of contamination in an aquifer progressively worsens the safety of our drinking water. Similar increases in contamination occur in lakes, rivers, streams, and oceans as we release more contaminants into our environment.

We can help reduce the resultant contamination. Consider how our individual personal actions cause contamination. We have already discussed the role of air, rain, and snow. Now add humans into the equation. As an example, go into your nearest bathroom, where you keep commonly used household personal care items. Pick up any toothpaste or mouth rinse. Read the labels and examine the list of ingredients. What do you find? For example, if your toothpastes and mouth rinses are similar to mine, many of the following strange-sounding substances will be present: hydrated silica, glycerin, sorbitol, PVM/MA copolymer, sodium hydroxide, propylene glycol. Remember that these personal hygiene items contain substances that we often deliberately put into our mouths, and they contain material that we would not knowingly swallow. Instead, we spit them out into our sink, after which they ultimately return as trace or micro-trace amounts to the tap water that we consume.

Pick up any hair shampoo, conditioner, styling product and you will seldom find any of the ingredients listed. Look at your laundry soap's ingredients, and they too will be generally obscure. Pesticides and herbicides used in our gardens and farmlands to kill pests and unwanted vegetation contain various toxic compounds; they are formulated to kill living things and are often carcinogenic chemicals. The gasoline, oil products, plastics (which contain volatile materials released by evaporation or sublimation into the air), solvents, and paints all contain myriad manmade chemical compounds. The composition of body-care products, such as lotions, cosmetics, salves, deodorants, antiperspirants, and make-up formulations are generally trade secrets, and the exact contents remain unidentified. Some of the remnants of body-care substances drain into ground water and sewage discharge as we shower and bathe ourselves.

Household cleaning and polishing substances are laden with manmade chemicals. They become part of the wastewater that we wash down a sink or throw out into our trash, which eventually ends up in a landfill. (The word "landfill" is a misnomer—the correct word is instead a "dump" where trash and waste containing many toxic and carcinogenic materials exists.) Food wrappings, plastic storage bags, and trash bags release chemicals into their contents and the air. Paper and plastic wrapping products and containers usually have manmade chemicals that leach into anything they contact, and they may evaporate into the air and condense on surfaces, allowing the cycle of water contamination to continue.

Take your own household inventory and you will realize that we are literally awash in a potential pool of contaminants that we *consciously* use yet *unconsciously* allow to enter our water supplies—and ultimately ourselves. The list of things that we use almost daily is the equivalent of a small chemical factory, whose output is about to become a part of our drinking water, our foods, and our bodies themselves. Remember, the vast majority of these contaminants are manmade substances that have been on Earth for seventy-five years or less. About seventy-five years ago, you were not exposed to these contaminants at birth. Today all of us, including embryos and fetuses developing in their mothers' wombs, newborns, and very young children, are exposed as they and their mothers use contaminated water.

Ultimately, whatever their origin, the contaminants often return to us in our water and food supplies, possibly in trace or micro-trace concentrations that can have devastating consequences after sufficient accumulation has occurred. Medical authorities will debate what is safe and what isn't. However, our health should not be the subject of a debating society. We must have meaningful investigative research. The requisite research is generally lacking for single chemical substances interacting with the cells in our bodies. It is totally lacking for the potential cocktails of contaminants that exist in our water and bodies today. Our bodies have become repositories for the toxic chemicals that contaminate our water supplies.

Biological research to determine the damage caused by chemical contaminants has an intrinsic impediment: humans cannot be the test subjects for obvious ethical reasons. Researchers, therefore, depend on laboratory animal testing. Many contaminant-induced diseases and cancers develop slowly, and a lengthy incubation period is required. Therefore, short-lived animals such as mice and rats may not manifest cellular changes, which require a long period of latency, in the lab. Other animals that are sometimes used, chimpanzees and apes, are more like humans, but they legitimately arouse ethical and moral objections. We are ultimately dependent upon the use of mice and rats for testing and cellular research into the biochemical effects of contaminants. Even such research is inadequate to determine the effects of chemical contamination on humans—because each of us possesses different susceptibilities to the biological damage caused by chemical contaminants—what, then, is an appropriate approach?

We cannot accept long-term clinical research of diseases caused by contaminants as the way out of this dilemma, because the damage, once detected in actual individuals, is out of the bag, so to speak, and will have likely already caused serious health problems in affected populations.

Therefore, what precautionary and preventive research can we perform to eliminate the commercialization of suspected disease-causing chemical substances before they are used in various products?

I believe that we must always err on the side of "cause no harm." The advances in cellular biochemistry now allow scientists to predict the probable cellular damage caused by chemical substances. Since all animal life shares similar cellular and sub-cellular molecular or atomic parts, all animal life will likely react to contamination in similar ways. We must perform biochemical and microbiological cellular research to determine the effects that chemical substances have on our "cellular" health. Let scientists, not politicians, government bureaucrats, or corporations, determine what is or is not safe. Let's get going and do this before it is too late. As recent scientific findings demonstrate, the time to act is now!

5

Recent Scientific Findings

The earth and all living things, including humans, are similar to guinea pigs used in a biological lab, yet our fate is impossible to predict because the environmental contamination variables to which we are exposed cannot be easily controlled or limited. We are on a treadmill of continual accumulation of manmade contaminants, increasing at an accelerated rate during the last three quarters of a century. *One thing is certain: we become more contaminated the longer we live, and innocent newborns are inheriting the contaminants from the living. This is not conjecture, but proven scientific fact.*

Several government and independent reports provide information and data about the extent of the ongoing contamination of our water supplies and our bodies. (Refer to reference sources listed in the Appendix.) The following sections contain summaries of some of these reports.

1999 U.S. Geological Survey

In 1999, an extensive report that assessed the quality of our nation's water supplies was published: "U.S. Geological Survey, 1999, The Quality of Our nation's Waters—Nutrients and Pesticides: U.S. Geological Survey Circular 1225." This report focused upon the contamination of ground water by herbicides, soil nutrients, and pesticides, including several that were no longer in use because of regulatory banning (DDT, banned in the early 1970s, and pesticides Chlordane and Aldrin, banned by the late 1980s but still in our water and soil). Sewage wastewater discharges were among the most significant sources of phosphorus and nitrates that contributed to the growth of algae and other aqueous plant life in streams, rivers, and lakes.

Algae, as those who live in or visit Door County in the summer know, is often abundant along the shores of Green Bay and Lake Michigan, where numerous sanitary wastewater effluent discharges loaded with phosphates and myriad other contaminants are located. Other significant sources of water contamination include various household, industrial, and natural human wastes and contaminants discharged from septic systems into lakes, rivers, and streams, particularly near the shorelines. Chemical contaminants also are in run-off from golf courses and private lawns that are treated with a variety of herbicides, pesticides, fungicides, and fertilizers. Various contaminants also flow off parking lots and streets as rains and snow wash them into nearby soils, creeks, rivers, lakes, and wetlands, or from sewer lines as they overflow. Roofs, sidewalks, and all similar impermeable surfaces shed contamination as these surfaces are "washed" by rain, melting snow, and winds. *Look closely at any sidewalk, street, or parking surface to get an indication of what is "washed" back into our sources of drinking water from these surfaces, as storm waters drain over them. You will be appalled as you see oil and fuel residues, cigarette butts, assortments of trash and waste, and all kinds of unidentifiable debris and residues.*

The contamination of domestic water supplies that comes from aquifers, reservoirs, and lakes is unavoidable as contaminants enter from various sources. Susceptibility to contamination varies according to the "openness" of the source. Reservoirs, lakes, and rivers, being naturally unprotected, are all very easily contaminated. Aquifers vary according to the geological nature and depth of the filtering soil and the abundance of surrounding wetlands that exist on and in the soil above them. Contaminants that enter our water sources may then return via our water tap.

The U.S. Geological Survey report cited above contains this particularly noteworthy paragraph:

> Mixtures of contaminants also require special consideration in assessing possible health and environmental effects, and thus in developing and improving water-quality standards. More than one-half of all stream samples contained five or more pesticides, and nearly one-quarter of groundwater samples contained two or more. These mixtures of pesticide parent compounds also occur with breakdown products and other contaminants, such as nitrate. Continued research is necessary to help reduce the current uncertainty in estimating risks from commonly occurring mixtures. As improved information the occurrence of contaminant mixtures should be considered when developing water-quality standards and monitoring requirements.

What has happened to improve water purity since the report in 1999? If any elected members of our Congress read *U.S. Geological Survey Report 1252* in 1999, what did they do to investigate the pesticide mixture question raised by the government-employed scientists? Did the politicians and political appointees, who run the agencies have the last word, do nothing? It appears likely, since there is a lack of evidence to the contrary.

Known Carcinogens and Tap Water

Many other reports are available to the public, to our elected officials, and to the news media. Some of the more significant reports are mentioned in the following paragraphs.

For the last eleven years, the National Institute of Environmental Health Sciences (a department within the U.S. Department of Health and Human Services) has issued a report to our Congress that describes connections between environmental contaminants and cancer. Both known and suspected carcinogenic substances are included.

These "Report on Carcinogens" (RoC) are seemingly lost in the bureaucracy of our federal government. *The latest RoC at the time of this writing, issued on January 31, 2005, included 148 mainly manmade chemical substances,* a miniscule fraction of the total of at least 400,000 in existence and the many thousands in common use. *Yet, it is significantly more than the 119 regulated chemical substances that the EPA and FDA include in water standards used by water utilities.* The Report on Carcinogens contains the following statement:

> The RoC does not present quantitative assessments of the risks of cancer associated with these substances. This listing of substances in the RoC only indicates a potential hazard and does not establish the exposure conditions that would pose cancer risks to individuals in their daily lives. Such formal risk assessments are the responsibility of the appropriate federal, state, and local health regulatory and research agencies.
>
> The cancer hazard that listed substances pose to any one person depends on many factors. Among these are the intrinsic carcinogenicity of the substance, the amount and duration of the substance, and an individual's susceptibility to the carcinogenic action of the substance. Because of these considerations, the RoC does not attempt to rank substances according to the relative cancer hazards they pose.

If this sounds like the typical political waffling that we are accustomed to, it probably is.

Clearly any alert member of Congress should have initiated follow-up actions to benefit the public health after receiving even one of these reports, let alone the eleven that have been issued to date. Why does our Congress ignore these reports? Why have the media ignored these reports? Why are the substances included in the evaluations so limited? Why do elected federal officials apparently ignore these reports even when they are submitted directly to them?

A consortium of organizations published another report on November 7, 2005, *Drinking Water Nitrate and Health: Recent Findings and Research Needs.* Participants included many international, university, and U.S. government agencies and departments, including the National Cancer Institute, National Institute of Health, and the University of Maastricht (Netherlands). Their investigation identifies nitrates in ground water and drinking water supplies and adverse health effects, including the "Blue Baby" syndrome (methamoglobinemia, caused by nitrate byproducts that interfere with the blood's capacity to carry oxygen). Ingested nitrates and byproducts contained in mother's blood can transfer to the fetus (through the umbilical cord), causing this life-threatening condition. Milk formulas that are prepared with water high in nitrates is another cause. Nitrates, as precursors of nitrogen compounds that form in humans (and other animals), are capable of causing deadly carcinogenic reactions. Nitrates in water are associated with childhood brain cancer, colon cancer in men and women, and prostate cancer.

A report was published in July 2005 by the U.S. Department of Health and Human Services, Centers for Disease Control and Prevention (CDC), entitled *Third National Report on Human Exposure to Environmental Chemicals.* This report (National Center for Environmental Health, Division of Laboratory Sciences, Publication No. 05-0570) is available as a CD ROM (*Third Annual Report on Human Exposure to Environmental Chemicals, 2005*) and contains data on 144 different chemicals (that were pre-selected for chemical analysis) in human blood and urine samples that entered the body from various types of environmental exposures. This report is limited in scope because of the very limited number of chemicals that were pre-selected for chemical analysis, but nonetheless the work represents an important baseline of human contamination caused by some environmental contaminants (i.e., manmade products). It is noteworthy that this government report also identifies many more toxic and carcinogenic chemicals than are currently regulated in our water supplies.

Many of the environmental contaminants included in the CDC report are identical to those found in tap water, reported in the EWG study cited below, and are known to cause serious diseases, including cancer. However, the sampling of analyzed chemical substances is woefully small and does not provide a sufficient assessment of the human body's contamination with toxic and carcinogenic substances. However, it is a good start, and maybe more information will be forthcoming.

Thankfully, dedicated scientists and technicians will do the necessary work if the politicians provide the funds, allow the results to become public immediately, and do what is required to implement the scientific findings for the public good. Most government-employed scientific and environmental personnel are inadequately budgeted by legislators to accomplish their needed work. Government-paid investigators who want to publish their research are sometimes prevented from doing so by their superiors, who are appointed, and by politically influenced administrators. Corporate-influenced legislators, elected state governors, and our elected federal government leaders (whose campaign funds are from corporate and executive "donations," and special-interest "contributions") are often complicit in these apparent obstructions.

The Environmental Working Group (EWG), a nonprofit research organization headquartered in Washington, D.C., published a report entitled, *A National Assessment of Tap Water Quality* on December 20, 2005. Their publicly and privately funded research showed that the tap water in the forty-two states included in the survey was contaminated with 260 different chemical substances. *Of the 260 chemical contaminants detected, 141 are unregulated, meaning that neither the EPA nor the FDA has established any safety standards for these contaminants.* The EWG investigation included the chemical analyses of over 22 million tap water samples from individual households where the community or municipality runs the water system. The database is available in an on-line format that enables individual searches to determine local tap water conditions throughout the United States.

Once again, many more toxic contaminants are in our water supplies than are regulated. Why are these additional known chemical substances not included in government standards? What are the safe concentrations, if any? What concentrations of various mixtures of the chemical contaminants are safe? All of these questions remain unanswered and unevaluated by our government.

Of particular interest in the EWG report is the ranking of states with the most significant pollution concerns. Wisconsin ranks second, following California. Arizona, Florida, North Carolina, Texas, New York, Nevada, Pennsylvania, and Illinois all follow as the top ten places where tap water is most contaminated. *The total number of people in the U.S. exposed to contaminants in our tap water exceeds 220 million, according to the EWG estimate.* This serious health problem deserves immediate governmental action, instead of the apparent indifference that we get from many of our elected politicians and their appointed government bureaucrats.

The contaminants found in tap water include those originating from industry, agriculture, general development, personal or household use, and water and sewage treatment processes. The limited data obtained in this study demonstrates that contamination of tap water is pervasive. More investigative work is required using ultra-sensitive analytical instrumentation capable of detecting very small traces of harmful chemicals and mixtures.

Another report, prepared by the Boston University School of Public Health and the Environmental Health Initiative of the University of Massachusetts titled *Environmental and Occupational Causes of Cancer: A Review of Scientific Literature*, published in September 2005, presents a comprehensive summary of medical investigations that correlate and implicate many environmental contaminants with specific and potentially deadly cancers. Included among the strong causal relationships are the following findings:

1. Arsenic as a cause of bladder, lung, and skin cancers

2. Chlorination byproducts in drinking water as a cause of bladder cancer

3. Asbestos and other fibers as a cause of larynx, lung, and stomach cancers

4. Petrochemicals and combustion products, including auto exhaust and volatile organic compounds, as causes of bladder, lung, and skin cancers

5. Pesticides as a cause of cancers of the brain, leukemia, and non-Hodgkin's lymphoma

6. Reactive chlorides, including vinyl chloride, as a cause of liver cancer and cancers of soft tissue

7. Metal working oils and mineral oils as causes of cancers of the bladder, larynx, nasal passages, rectum, skin, and stomach

8. Solvents and other petrochemical byproducts, such as benzene, as the causes of leukemia, non-Hodgkins lymphoma, kidney, and liver cancers, respectively.

Suspected but weaker causal relationships, too numerous to list, are strong indicators of the breadth of the human impact via environmental contamination.

We all know that many human and animal diseases come from infectious viral or bacterial sources. Others are due to genetic factors. However, the toll due to environmental contamination is significant and increasing, as childhood cancer deaths indicate. Nearly one-half of the U.S. population will contract some form of cancer during their lifetime. New treatments are extending the duration of survival, but cancer is now second behind heart disease as the major cause of all death. Environmental causation seems highly probable in a large percentage of these cases. We must err on the side of caution and constantly work to reduce and eliminate the toxic and carcinogenic contaminants in our environment.

The references cited in the Appendix of this book, as well as many other reports and worldwide findings, verify the obvious. Contaminants in our environment that we consume, inhale, ingest and contact become constituents in our blood, vital organs, skin, bones, and tissues and circulate throughout our bodies. Our tap water is a major source of toxic and carcinogenic contaminants.

Every living thing is subject to man's apparent manic desire to produce and use substances without regard to the health consequences. Our bodies become more contaminated, as does our environment (often through our own actions). No one can predict the individual day of reckoning when toxic and/or carcinogenic chemical contaminants in our body reach accumulative concentrations that singly, or in combination, causes illness, premature disability, or death. Some species of animal and plant life have probably already become extinct because of manmade toxins and carcinogens. *Who, and what, will become extinct because of toxic contamination of earth's water, air, and soils?*

Coal Storage Area Adjacent to a Wetland (Ground Water Contamination
Source)

Oil Storage Tanks Adjacent to a Wetland (Ground Water Contamination Source)

View of a Bulkhead Where Treated Sewage Effluent Enters Lake Michigan
Near the Fox River Estuary (Abandoned Dredge Line and Power Plant in
View Also)

Island Depository in Lake Michigan for Dredged Lake Materials
Containing PCB's

Large Coal Storage Area Adjacent to the PCB Contaminated Fox River in
Northeastern Wisconsin

Incinerated Sludge Ash from a Sewage Treatment Plant Disposed of in a Landfill (Ground Water Contamination Source)

Woodlands and Farm Lands Being Transformed into a Four-Lane
Highway to and from Door County, Wisconsin, a Tourist Area Bounded
by Lake Michigan

Sludge Thickening Tanks for Solid Sewage Wastes Being Prepared for Incineration and Residual Ash Landfill (Courtesy of Green Bay Municipal Sewerage District)

Manicured Golf Course in Peninsula State Park Adjacent to Lake
Michigan Waters (Chemicals Used for Maintenance are Water
Contamination Sources)

A Relatively Untouched Lake Michigan Shoreline in Newport State Park,
Wisconsin

Dense Condo Development Adjacent to a Lake Michigan Shoreline
(Water Contamination Sources)

Asphalt and Parking Areas Adjacent to a Lake Michigan Shoreline (Water Contamination Sources)

6

Will Government Protect Our Environment and Water Quality?

The Environmental Protection Agency (EPA) was the outcome of an executive order that President Nixon made in 1970. It is now the only federal government agency with the responsibility to formulate and enforce regulations to protect the American people from potential harm caused by environmental degradation. Before then, the individual states were the only government entities that could enforce environmental controls, although they seldom did.

The EPA reports directly to the President. Eleven appointed administrators have managed it since its inception. They were all political appointees selected by the President in office at the time. Only *one* of these eleven was in part qualified, by virtue of formal training in a relevant scientific field. Most were lawyers. The current administrator, however—Stephan L. Johnson, appointed in 2005 by President George W. Bush—worked his way through the ranks in the EPA and has relevant scientific training (an M.S. degree in biology). All of the appointees are political, and as such serve as long as they abide by the policies established by the current President's administration, or they decide that they prefer greener pastures elsewhere. Some have likely departed for their moral beliefs, citing "personal reasons."

The annual operating budget of the EPA is part of the overall federal budget submitted by the President to Congress annually. This procedure applies to all federal departments and operations, in theory. The budget formulation and approval process is modulated by political trade-offs and "pork-barrel" spending (otherwise known as "earmarks"). Every representative and senator wants a piece of the budget pie for the districts or the states that they represent. Special lobbying groups receive largesse from lobbyists for favors that they have provided to their supportive politicians. This type of behavior has been well publicized recently after numerous politicians were exposed and even convicted of crimes.

Even at best, the result is often a hodgepodge of approved expenditures that satisfy specific desires but not always needs. At worst, the resultant budget is so restrictive that the EPA workers cannot accomplish what is actually required to protect the environment. The President and the political party in power often seem to have other priorities that trump the needs of the EPA. External influence often drives priorities. These influences can be either positive and benefit the people or negative and benefit special interests, usually corporate. The latter seems to prevail. In the past, we the people have sometimes benefited from administrations that were environmentally friendly. President Teddy Roosevelt was an avid environmentalist, and more recently the Clinton/Gore Administration generally fostered environmental improvements.

In contrast, The Bush/Cheney Administration has consistently resisted environmental improvements and has in fact negated some of the previous Clinton/Gore actions. The Bush/Cheney Administration and their appointed EPA administrator weakened environmental regulations adopted by the Clinton/Gore Administration (to reduce toxic mercury emissions from coal-fired power generating plants). Our nation's promise to participate in the Kyoto Treaty to reduce global warming emissions was reneged on by the Bush/Cheney forces. Robert F. Kennedy, Jr. 's book *Crimes against Nature*, published in 2004, details many anti-environment actions that have been taken by the Bush/Cheney Administration designed to protect corporate interests at the expense of public health.

The assault on the environment by the Bush/Cheney Administration continues unabated. The inaugural meeting of the Asia-Pacific Clean Development and Climate Partnership was held in mid-January 2006 in Sydney, Australia, for the purpose of arriving at an agreement on the methods to combat global warming, particularly by limiting greenhouse gas emissions. The meeting was attended by the foreign ministers of the six partner nations: Japan, the U.S., Australia, India, Korea, and China, which represent more than half of the world's economy, population, and energy use. Australia and the U.S., non-signatories of the Kyoto Treaty, established this group to create an alternative to the Kyoto Treaty whereby business instead of government would determine the remedies. The participants included 60 round-table participants and 120 business observers. *Representatives from environmental groups were not included, and the media were prevented entry at the sessions.*

The Bush/Cheney Administration selected Dr. Samuel Bodman, secretary of the department of energy, to lead the U.S. delegation. Interestingly, the current and qualified administrator of the EPA did not attend. According to "leaks," Dr. Bodman stated in a speech he gave, "The people who run the private sector—they too have children, they too have grandchildren, and they too live and breathe in the world and they would like things dealt with effectively, and that's what this [meeting] is all about." This encapsulates the viewpoint of the Bush/Cheney Administration—*big business knows best* and *will do what is right*. Our world's environment would be so much healthier if this were true. However, approximately seventy-five years of corporate-caused and politically endorsed, contamination, largely without regard for the environmental and health consequences, demonstrates that this "mantra" is false. Corporations cannot, or will not, curtail their own benefits by taking any actions that substantially reduce profitability or executive bonuses (awarded when profits are high). Seventy-five years of corporate-caused pollution have proven this indisputably. Thirty-five years of EPA regulations, surveillance, and actions may have slowed the flow of contaminants slightly, but the onslaught on our environment by private corporate polluters continues nearly unabated. This results from an insidious linkage between corporate interests and governments—between lobbyists and the local, state, and federal governments.

Individuals working their way up the corporate ladder as they aspire for more power and money realize that they must remain in step with corporate objectives to achieve their desired promotion. Consequently, corporate cultures generally breed and reward individuals who do whatever is required to increase the profit stream. Anyone who interferes with this prime purpose is not welcome and will usually not become a member of the top executive cadre. Sometimes public and privately owned businesses create highly ethical creeds that do benefit the community at large. These situations are rare, and they are overwhelmed in number and influence by the majority of corporations, which are motivated by profit only. Will the "private sector" ever be the protector of our environment, as Dr. Bodman and the current Bush/Cheney administration believe? This is as likely as reconstructing a broken egg.

The establishment of a meaningful environmental protection program requires independent creation and enforcement of regulations based on science and technology. Policies to eliminate dangerous contamination must be the first line of defense. Preserving and protecting people's health, not corporate profits, must be the only goal of an effective environmental protection program. The political apparatus that awards politicians with payoffs (called campaign contributions and political action committees, PACs, promotions) as well as junkets by corporations via their lobbyists subvert not only the EPA but virtually all forms of local, state, and federal government.

The EPA, the Department of Energy, the National Institute of Health, and virtually every bureaucratic organization within government have become tools of business and business lobbyists, and this fox-in-the henhouse approach has achieved exactly what the Bush Administration intended. "Environmental protection," is merely a sound byte, along with "clean air" and "clean water." These words echo through the chambers of Washington, D.C., as if the echo itself were the answer. Indeed, to many politicians, it is! Sadly, individual state environmental agencies are often beholden to business influences, just as those anchored in D.C. are. Corruption is rampant, and serious environmental regulations and action are lacking.

Our Federal government and the EPA cannot protect our environment unless the fundamental basis of action changes from *business protection* to *people protection*. Anything short of this will lead to the inevitable worsening of environmental contamination and the destruction of life. Our water, air, and global climate are all at stake, and they depend upon the right political decisions.

7

Not in My Back Yard?

You may still be wondering how all of this applies to your daily life. Why should you be concerned? Nothing has ever happened in your neighborhood to justify questioning your water quality. The purpose of this chapter is to persuade you that contaminated water is everyone's problem and is in everybody's neighborhood. The not-in-my-backyard mentality will not protect us from the pervasive problem of water contamination and the health impairment that prevails around us.

Let's look at some recent happenings to crystallize this point. The material is so abundant that it is difficult to summarize. I have decided to present some of the information that I collected during the months of February through April 2006. The categories by no means encompass all the areas of water contamination, which are in fact ubiquitous, defying compartmentalization. Are any of these in your back yard?

Benzene in Soft Drinks

The previously cited Environmental Working Group recently released a report (April 2006) stating that the Food and Drug Administration knew for three years that some soft drinks had unsafe levels of carcinogenic benzene, yet failed to notify the public. Benzene is not in the product as produced, but it can form due to a reaction that occurs between certain ingredients used to preserve and flavor the drinks. Beverage producers are now reformulating the ingredients to avoid the reaction. However, for several years, millions of us swallowed benzene in soft drinks without our knowledge. Why did the FDA withhold the information for at least three years? How much illness and cancer resulted from this unforgivable and inexplicable governmental behavior?

Local Cancer Outbreaks

In the township of Dover, New Jersey, a citizens group has been attempting to protect their children and themselves from outbreaks of childhood cancer for over ten years. A rare form of leukemia, brain cancers, and central nervous system cancers were affecting a "higher-than-normal"(300% greater) number of children in this community of 95,000 people. The local water company, United Water, failed to notify the residents that their water contained radioactive and other toxic chemicals that exceeded regulated allowable limits, and they were fined for this delinquency. Two local activist groups, Citizens Action Committee on Childhood Cancer Cluster and Toxic Environment Affects Children's Health (TEACH), have been trying to get local governments to protect their water supplies for a number of years, and they are still at it, as unusual cancers are occurring in their children. The sources of some of the contaminants were previous industrial sites in the community. The companies discharged toxic chemicals that leached into the ground water and the aquifer that serves the community and that the local water company supplies.

On April 25, 2006, Massachusetts state health officials advised residents of the community of Ashland to consult their doctors about possible cancer risks associated with exposure to contaminated water in a wetland that existed about twenty years ago. An article written by Beth Daley (*www.boston.com*) described the results of a seven-year study by the state, which showed that individuals who had contacted the water in the contaminated wetland were at greater risk for cancer. The wetlands which had been contaminated with about one hundred toxic chemicals (used to make dyes from a previous industrial operation that was discontinued in 1978) were used by children who played in the multi-colored wetlands. The area is now "sealed" by a lining after the EPA tried to remove the contaminants, although they admit that more cleanup work is required.

The state of Massachusetts study revealed that seventy-three individuals (now adults) exposed to the contaminated wetlands when they were children at play have contracted some form of cancer before age thirty-five. Some young adults who were exposed died before the age of twenty-five from rare forms of cancer. The state advisory recommends that all individuals who were exposed see a doctor about possible cancer risks.

Contaminants in Fish

Our oceans, rivers, and lakes are the habitat for fish, which are a major source of our food and the food of natural predators (including other fish and predatory birds). As our waters become increasingly contaminated, so do the fish and the other life forms that live in or use the contaminated waters. Fish contamination is a worldwide problem. In April 2006, the Maryland Public Interest Group issued a report on mercury levels in fish and shellfish caught in Maryland. The report, based on data gathered by state agencies, showed that 59% of the samples analyzed (1,939 fish) had mercury levels that exceeded recommended maximums, while about 9% had mercury levels that were ten times more than the recommended maximums. The mercury sources found in the waters of Maryland are mainly from coal-burning power plant emissions. These form a compound that enters fish as they swim in contaminated waters. Mercury is toxic to all water life and all who consume it, causing neuromuscular and brain damage that can start in the unborn fetus if the mother has consumed mercury-containing fish.

Mercury, pesticides, and PCBs form a toxic cocktail that has caused serious disease among the sturgeon in Columbia River Reservoirs. These gargantuan fish, which live in certain fresh waters, contain toxic chemicals, as do the people who consume sturgeon. Researchers have reported that pesticides, mercury, and PCBs are found in many sturgeon; these chemicals impair their ability to reproduce and spawn in addition to creating a toxic fish dinner. A new study by Pacific University will attempt to quantify the sturgeon contamination in the severely contaminated Columbia River reservoir water.

The oceans are the world's largest bodies of water, providing most of our fish. The ocean waters now contain complex mixtures of toxic and carcinogenic chemical substances that are endangering all sea life. In addition, artificial fish farms that confine and breed fish for commercial purposes in ocean cages are increasing the concentration of contaminants in the fish produced for human consumption. A recent report by the Scottish Environmental Protection Agency, described in Scotland's *Sunday Times* in April 2006, decried the practice of fish farming, in which as many as 50,000 fish are crowded into submerged cages along the Scottish ocean coastlines. The report stated, "Toxic chemicals are used in the production of farmed salmon and can be present in the salmon sold to consumers." Antibiotics used to prevent infestation of salmon by sea lice and other contaminants located near shorelines are concentrated in the salmon grown there. Fish farms are part of a very fast growing industry that poses a serious threat to consumers if contamination prevails.

Scientists at the University of California at Santa Cruz, who measured the levels of PCBs and DDT in two species of albatross that fly over vast ocean areas, seeking their fish dinners recently released a report about the consequences. The report, published on April 6, 2006, in *Ecological Applications*, revealed that DDT, PCBs, and mercury measured in the blood of certain albatross species was two to four times greater than that measured ten years ago. They conclude that the open ocean waters and fish that albatross consume contain more contaminants now than they did ten years ago. (The toxic and carcinogenic pollutants are, of course, manmade).

The diminished fish populations have roused the attention of some members of Congress, who are worried about the commercial interests and are considering legislation to regulate fishing practices. The supply of fish off U.S. coastal waters has declined during the last several years to the point where fish stocks are very limited. Excessive fishing often is the reason given for the diminished supply. It is undoubtedly a factor. However, when will politicians wake up and realize that water contamination is another reason, and possibly the main one, that the supply of edible fish is dwindling?

Legal Actions

When corporations and governments fail to protect the citizens from harmful contamination, legal actions are required to force the contaminators to cease, and to clean up the toxic messes that they created or allowed. Because of the all-too-frequent resistance of governments and corporations to protect the public from toxic contamination, the legal system is being used more than ever before to protect our environment and water quality. Several recent examples are included in this section to illustrate this trend.

Erin Brokovich started the ball rolling, but corporations continued to deny the health problems caused by chromium. After years of delaying tactics (including the withholding of internal corporate industry data) and denial by industry lobbyists, the Occupational Safety and Health Administration is about to impose stricter limits on worker exposure to this known carcinogen. This government agency has been resisting the tighter limits for over thirty-five years, but a federal court order required them to implement the new regulation in 2006.

The New Jersey Supreme Court sided with the state and upheld New Jersey's strict new standards for the decontamination of polluted water in or under industrial sites. The ruling, made on February 28, 2006, overcame the legal challenge made by Federal Pacific Electric Company, the state Chamber of Commerce, and various organizations representing chemical and petroleum industries. Many state activists, environmental organizations, and the Rutgers University Environmental Law Clinic succeeded in overcoming the self-interested corporations and have forced them to decontaminate groundwater sources (aquifers) before any new development can take place.

The state of New York has sued the EPA to obtain records that the EPA has refused to release. These identify the volatile organic compounds found in common household and industrial products, including paints. The state of New York has been trying to get the records from the EPA for the last two years to provide a scientific basis for establishing clean air regulations. (Air quality has a direct impact on water quality, as discussed earlier.) Corporations that produce the materials claim that the information constitutes trade secrets, and the EPA seems to agree. A prominent Ohio senator, George Voinovich, has worked on behalf of a major paint company headquartered in Ohio to bolster the EPA's position. Another state, California, has also attempted to overcome the EPA refusal to create stricter environmental standards than the federal government and industry are willing to allow. Meanwhile, the New York federal lawsuit filed in February 2006 will be an important case in determining whether the people or corporate interests will prevail when it comes to environmental issues.

Another important ruling was made by a Federal Appeals court on March 17, 2006. The three-judge Appeals Court in Washington, D.C., unanimously ruled that the EPA violated the Clean Air Act when in 2003 they allowed industry to make equipment changes without environmental review. The Bush Administration had weakened regulations on behalf of corporate interests that wanted to avoid additional environmental control costs. The issue dealt with emissions from coal-fired power plants, factories, refineries, and chemical plants that produce many millions of tons of pollutants. This ruling is a temporary win for the public, but industry lobbyists are on the move to get Congress to rewrite the Clean Air Act and allow them to contaminate the environment as they desire.

The legal battles will continue as environmental advocacy groups representing public interests battle corporations, the Bush Administration EPA, and other agency appointees who want to weaken environmental regulations for the benefit of corporate profits. Two very important legal cases are now under review in the U.S. Supreme Court, and the decisions that the newly constituted Supreme Court makes may establish the immediate future of environmental regulatory actions. Two new justices, Associate Justice Alioto and Chief Justice Roberts, could decide the future of our water quality. The two cases involve the question of EPA and the Corps of Engineers' jurisdiction over wetlands and navigable waters. If the developers prevail and the Supreme Court reduces the scope of protection for wetlands and navigable waterways, we will see an avalanche of development on and adjacent to wetlands that laws presently exclude from development. The consequences of reversing these long-standing protections would be devastating to our water quality throughout our country. Wetlands create a natural buffer and a continuous filtering mechanism that helps shield our aquifers from contaminants.

Arsenic

According to the EPA's own data, approximately 10 million Americans are drinking water that contains unsafe concentrations of arsenic. Arsenic, a naturally occurring chemical element that is unearthed during mining operations, can saturate ground waters and the ambient air and cause water contamination. Some rocky terrain in Utah has very high natural arsenic concentrations; these contaminate water supplies and cause disease, cancer, and death. Other states, including some areas of Wisconsin, Maryland and Virginia, have similar arsenic problems. Would you knowingly swallow doses of arsenic with every sip of water that you drank? I think not. Are you surprised to find out that if you live in a rural community with a population of less than 10,000, you might be "allowed" to drink water containing excessive arsenic? The current EPA maximum concentration limit (MCL) for arsenic is 10 parts per billion, a concentration level formulated during the Clinton administration but not enforced. The Bush EPA rescinded the standard, but Congress reinstated it. Currently, the EPA is planning to raise the limit to 30 ppb for municipal water systems with fewer than 10,000 people because of the added cost of water treatment. The National Rural Water Association (a lobbying group representing rural water treatment operators) is opposed to the more restrictive standard. Whom do they really represent? If the EPA succeeds in raising the MCL, the looser standard could become common across the country and increase the overall contaminant levels of arsenic in our drinking water. Stay tuned to EPA on the Web to see how much arsenic they will allow you to drink with each sip of water from your tap.

The natural presence of arsenic in some terrain represents a serious water contamination problem. Contamination is spread as a consequence of mining operations, which expose arsenic-containing minerals to the atmosphere to be dispersed by winds and storm waters into otherwise arsenic-free regions nearby. This type of contamination has occurred near mining operations in the community of Deer Lodge, Montana. The residents there have been trying to alleviate the resultant water contamination for about twenty years, to no avail.

An article written by Perry Backus was published in *missoulian.com*, on April 9, 2006, that describes arsenic contamination of the Upper Clark River area, where more than 80 tons of arsenic was dispersed into the environment daily as vast copper mining and smelting operations took place. Most of the mining operations have ceased, but the toxic remains contaminated the earth and ground waters, as well as in the water used in the surrounding communities that border 120 miles of the Clark Fork River. When the mining operations were thriving, the residents merely shrugged and accepted the contamination as an unfortunate side effect of economic advantage. Now the economic advantage is gone, but the arsenic remains, poisoning the recreational and well waters of the area. The EPA put the highly contaminated area on the Superfund for decontamination, but until money is available for the massive cleanup work, nothing will rid the area of the toxins. The needed decontamination has been on the Superfund's hot list for twenty years. The mining companies who caused the contamination are now part of two huge petrochemical corporations, Atlantic Richfield and British Petroleum. The affected communities are now trying to obtain the many hundreds of millions of dollars required from their deep pockets.

Nitrates

Do you live near a factory farm or natural composting operation? If so, you should check the nitrate content of your well water. The *Missoula Independent* weekly journal of April 13, 2006, tells a story that is relevant to many areas. If you live in Western Montana near the West Valley, you probably already know that nitrates probably contaminate your well water. The open fields in the area are disposal areas for dairy cattle waste and nitrate-containing fertilizers. The shallow aquifer that serves the wells was contaminated with nitrates from these sources. Some well water concentrations have exceeded 50 milligrams per liter of water, enough to cause the death of a child within weeks. (Nitrates are converted to nitrites in our bodies and can cause cancer and Blue Baby Syndrome in infants and children.) Deeper wells were bored, and now the contaminated water is working its way into deeper aquifers and out of the deeper wells. This example illustrates a wide-ranging problem that is associated with factory farming operations and the indiscriminate use of nitrate-rich fertilizers. Sewage-treatment operations and septic tanks also release nitrates into the soil or into effluent streams later discharged into lakes and streams. Wherever you live, have your water checked for nitrates and bacteria at least annually.

Your Personal-Care Cabinet

A Canadian newspaper, the *Ottawa Citizen* (Ontario), published an article on March 5, 2006 that dealt with the indiscriminate use of infant and childcare products containing unsafe ingredients or substances that have unknown effects upon the health of children and adults. Adults can choose what to use or not, but children generally cannot. The thrust of the article, written by Shelly Page and Susan Allen, is that we know very little about the safety or danger of the large number of personal-care products that we use and dispose of in substantial quantities. The waste from these products will contaminate your drinking water. Their article cites an Environmental Working Group action that prompted the U.S. Food and Drug Administration to issue a warning last June 2005 that stated, "Manufacturers using untested ingredients may soon be forced to tell consumers with product labels that say, 'Warning: The Safety of this product has not been determined.'" Read the article on the Internet that can be accessed at www.canada.com. When selecting your personal-care products and those of your children, ask the following question: "What do I know about the safety of the ingredients in this product?"

Another Canadian report, by the Canadian Institute for Environmental Law and Policy, reviewed data obtained in North America and Europe suggesting that several hundred medical drug traces and cosmetic residues are abundant in many waterways. The report stated, "Pharmaceuticals and other emerging contaminants are widely present in the streams, lakes, and rivers and groundwater in densely populated regions of the country." A summary of the findings was presented by Martin Middlestdt on March 6, 2006, in the on-line version of the *Globe and Mail* (www.theglobeandmail.com). His article went on to say that 50 to 90% of the active ingredients in pharmaceutical drugs are not absorbed in humans or animals and are expelled from our bodies (into our toilets and into sewage systems). The human sewage is treated and discharged as treated liquid sewage effluent (contaminated water) into rivers, lakes, or in sludge landfills. Individual septic systems slowly release residues that can gradually migrate back into aquifers or other water sources and back into our drinking water. Some residues will be adsorbed or absorbed into soils and be retained rather than migrate. This is one indispensable value of vacant, undeveloped land. The EWG study

(Reference 1, Appendix) described in Chapter 5 of this book verifies the cycle of contamination described above.

Vietnam Agent Orange Contamination

The Vietnam War was a very sad chapter of recent American history. That war caused as much soul searching as any war in our history. One aspect of the war that is seldom discussed or even realized pertains to the water contamination that still exists in Vietnam thirty-five years after the war ended. Many thousands of our veterans who fought in Vietnam were also exposed (albeit for a much shorter duration than the Vietnamese people) to the toxic contamination caused by the use of Agent Orange (a dioxin-laden herbicide). The on-line edition of the British paper *The Independent* published an article on April 1, 2006, that describes the terrible legacy of that war. Much of the land defoliated with Agent Orange is still contaminated, and the water there is too toxic to use. If it is used, often when there are no other choices, serious or fatal diseases result. Terrible birth defects and abnormalities ranging from chronic deformities, spina bifida (a life-threatening spinal deformity characterized by the contamination and leakage of spinal fluid), chronic neuromuscular diseases, and cancers are now afflicting over 500,000 children, who will suffer for all their lives with the effects of the toxic and carcinogenic dioxin contamination. Last year, Vietnamese war veterans sued Dow and Monsanto corporations, claiming they knew that Agent Orange was toxic and potentially deadly, and that it was much more than a defoliant to eliminate bush and tree coverage. The Vietnam veterans lost their case, but they are appealing. Another civil lawsuit to be tried in a U.S. court this year involves whether the civilian Vietnam victims are entitled to some financial compensation. However, there can never be adequate compensation for the toxic contamination of their water, which has caused more than a generation of suffering, death, and despair.

Our own veterans sued the seven manufacturers of Agent Orange and received $180 million in a class-action suit settled over twenty years ago. The completely innocent Vietnam victims of this "defoliant" attack are the children poisoned by their water supplies where they happened to be born. This is one of the worst environmental disasters ever—unleashed on an innocent civilian population, Vietnam military forces, and our own U.S. fighting troops. Just like the Vietnamese people, large numbers of our veterans are now undergoing treatments for the lasting damage to their health caused by Agent Orange. Wars are always hell, but the use of Agent Orange is an example of monumental governmental/corporate duplicity and indifference, and it violates fundamental human rights. Keep informed as these future legal cases unfold, and speak out for justice.

Landfill Contamination

One of the most insidious types of contamination is that caused by the leaching of toxins from landfills. Landfills are toxic storage dumps for waste materials, or "toxic dumps." They are sites to dispose of wastes and store them for an indefinite duration. The toxic contents of landfills gradually leach into the soil and diffuse into ground water and aquifers. There they make a return trip to your water tap. Therefore, you must be concerned about all so-called landfills, since they are all sources of drinking water contamination.

The residents of the Saginaw Bay area of Michigan have recently become suspicious of toxic chemicals and radioactive waste that is emanating from a nearby landfill and infiltrating a wetland in a recreational area. An on-line article in *mlive.com* ("Everything Michigan"), written by Jeff Kart on March 4, 2006, describes how local residents have taken action to make the area off limits to schoolchildren, who had previously gone to the wetland area for field studies and recreation. The wetland surface water drains into Saginaw Bay and into the aquifers that provide drinking water. The landfill leaches lead, arsenic, mercury, and chromium, and low-level radioactive substances from an industrial landfill established by a local company many years ago, which has now been inoperative for about twenty years. The cement operations ceased, but the contamination continues as the toxic chemicals leach into the ground and lake waters, used by visitors and residents alike for recreation, drinking, cooking, and bathing. Visitors to this resort area may unknowingly swallow water-borne toxins when they visit.

In the resort city of the Cape May, New Jersey, the local Sanitary Township Landfill is discharging toxic contaminants into the aquifer that provides drinking water to residents supplied by the water utility company. An article written by Jack Fichter in the *Cape May County Herald* on March 29, 2006, describes the seriousness of the toxic contamination of residential water supplies. Many unregulated as well as regulated toxic and carcinogenic contaminants were detected in household tap water, including volatile organic compounds, pesticides, and numerous other toxic contaminants, many exceeding MCLs for drinking water by significant amounts (thousands of times greater than what is considered safe). The county has initiated a suit against the former owners and operators of the sewage landfill, and affected residents are resorting to individual home-based purification systems to reduce the contamination. How many sewage landfills contain toxic and carcinogenic sludge that is gradually leaching into the aquifers that provide your drinking water? Check out your local sewage treatment plant practices and seek impartial information.

The resort area of Little Traverse Bay, near Petoskey, Michigan, is known for the beautiful, rich brown and white—veined fossilized coral, known as Petoskey stones, that at one time covered the nearby Lake Michigan beaches. The stones have decreased, but now an old and unwelcome visitor has come, in the form of toxic water contamination. In the recently developed resort of Bay Harbor, over-looking Lake Michigan beaches, many very upscale property owners have realized that their drinking water is now contaminated with many toxic chemicals that have leached (from a former industrial landfill) into the aquifer that supplies their water. An article written by John Flesher in the April 8, 2006, issue of *det-news.com* describes the situation that has been brewing for years. The brown field (a term used for fallow and previously used industrial land) that became the exclusive 550-home Bay Harbor development is now experiencing the conse-quences of years of contamination caused by a cement company that stopped operations in 1981. The landfill contained high concentrations of toxic mineral wastes, which have gradually leached into the aquifer and into Traverse Bay. The developers of the plush residential area have committed to clean up the contami-nation and toxic residues from the landfill. However, these should have been removed many years ago, and they should not have been allowed to accumulate in the first place. How many toxic industrial landfills exist throughout the U.S. that are unknowingly contaminating your water supplies? Research what existed in your neighborhood before development transformed your area into housing developments, shopping centers, parks, schools, and entertainment centers. Are old landfills adding to the contamination of your water? If so, you should take action to remedy the situation to protect yourself and your family.

The penitentiary in your locale could be a major source of landfill contamina-tion, since many prison inmates are used as cheap labor to process electronic devices for recycling and, ultimately, waste disposal in landfills. This generally unknown facet of waste processing is the subject of an article entitled "Toxic Recycling," written by Elizabeth Grossman in the November 21, 2005, issue of *The Nation* magazine. The article describes how a division of Federal Prisons Industries (known as UNICOR) employs prison inmates to disassemble dis-carded electronic devices, after which the useless remnants are stored in landfills. These remnants often contain toxins, including cadmium, mercury, lead, various plastics, and carcinogenic organic chemicals. The article states that this waste dis-posal creates about 5 to 7 million tons of waste per year and is one of the most rapidly increasing sources of industrial toxic waste filling municipal landfills throughout our country. The highly toxic build-up of these in landfills threatens the water quality everywhere such a landfill exists. In addition, the air contamination in prisons where these operations are located poses a threat to the health of prisoners and guards alike, contaminating the air in the surrounding community as well. Check the landfills near you to determine if these very toxic wastes are included and if they present the potential for water contamination. Do not allow the slow leaching of these contaminants to poison your drinking water.

Industrial and municipal waste and sewage landfills represent potential sites for substantial ground and surface water contamination, which ends up in the water we drink. The wastes may linger in landfills for years before they begin to leach into the surrounding soil and into water supplies, as storm waters saturate the wastes and accelerate the leaching action. Once contaminants reach an aquifer, they will spread; creating plumes of contamination that can affect very large areas. They can also directly contaminate rivers, lakes, reservoirs, and streams. Tracing the source of contamination after this occurs may be very difficult. Private wells are very vulnerable to the sudden and unexpected contamination by plumes of contaminants that start to spread within an aquifer. Municipal or private treatment facilities may spot such occurrences quicker than individual homeowners are able to; because water utilities are required by state and federal regulations to monitor treated water for certain regulated chemical substances. Unregulated chemical contamination may or may not be detected depending upon the specific practices of the local water treatment plant and the type of contaminant that may unexpectedly appear. Check with your local water utility company to determine their practices, the frequency with which they have the water analyzed, and the contaminants they look for. Obtain copies of the annual reports that they are required to publish and review them diligently to assure yourself that your water meets you expectations. *However, remember that hundreds of unregulated chemicals that may be present in our drinking water are not monitored or controlled, and need not be reported, by our water utility company.* Singly, or in combination, they pose health risks, as do the 119 regulated substances (which are regulated because they are acknowledged to be toxic and/or carcinogenic) even if they are present at concentrations less than the designated MCL for drinking water.

Drug and Cosmetic Residues in Water/Sewage Effluent

A recent report by the Canadian Institute for Environmental Law and Policy reviewed data obtained in North America and Europe that disclosed that traces of several hundred medical drugs and cosmetic residues are abundant in many waterways. The report stated, "Pharmaceuticals and other emerging contaminants are widely present in streams, lakes, rivers, and groundwater in densely populated regions of the country." A summary of the findings was presented by Martin Middlestadt on March 6, 2006, in the on-line version of the *Globe and Mail* (www.theglobeandmail.com). His article went on to report that 50 to 90% of the active ingredients in pharmaceutical drugs are not absorbed by humans or animals, and are excreted into sewage systems or waste disposal. Human sewage is eventually discharged from sewerage treatment plants (as treated sewage effluent) into rivers, lakes, or sludge or into septic systems (from which liquids seep into the soil and solids are emptied and taken to treatment plants). The residues can gradually migrate back into our aquifers or other water sources, and into our drinking water. The EWG study (Reference 1 in the Appendix, described in Chapter 5 of this book, verifies the cycle of contamination described above.)

I recently visited a large municipal sewage treatment plant, where I had a very informative discussion with the executive director. We discussed the contamination described above, and he acknowledged that the current sewage treatment technology is incapable of removing these contaminants as well as other toxic substances. Most of the treated waste from typical sewage treatment plants flows into lakes, rivers, and oceans, where residual contaminants flow as well. Subsequently the sludge is incinerated and the toxic ash is stored in landfills. A typical sewage treatment plant that serves a 200,000-people population discharges 35 to 50 million gallons of treated wastewater a day into the body of water that it discharges into. The toxic, carcinogenic residues, not removed by treatments, end up contaminating the waters that we use daily, as the aquifers and reservoirs are recharged by groundwater or storm water. Our drinking water supplies are continuously subjected to these contaminants, since drinking water treatments do not subsequently remove them either. Most of the chemical substances originating from drugs and personal-care products are not among the regulated drinking water chemicals. (They are not reported by municipal drinking water plants.) They pass through the sewage treatment processes, which was not designed to remove them, and into the effluent stream or sludge, from which they contaminate our drinking water sources. Sewage disposal represents a major threat to drinking water quality throughout our country.

Pesticide Mixtures

The on-line publication *www.ourstolenfuture.org* reviewed the results of a recent investigation undertaken by researchers to determine how combinations of pesticide residues affect tadpole and frog reproduction. Using the same pesticides employed in Nebraska's agricultural fields to control pests, they discovered that very low concentrations of various mixtures caused the death of frogs and extinction of tadpoles. The work, scheduled for publication in *Environmental Health Perspectives* by T. B. Hayes et al, is due out during 2006. The report shows that tadpoles suffered heavy mortality when exposed to a mixture of nine different pesticides, even though the exposure to any one of the pesticides alone caused low mortality. Frogs have generally disappeared from habitats where they once were abundant, and the incidents of malformed frogs have been reported throughout the world. These extinctions and deformities are probably the result of widespread chemical contamination of water that frogs live in. Unusual fungal and bacterial diseases were associated with the inflicted tadpoles. The significance of this limited study is that we humans cannot rely upon MCLs for water standards, since they consider one contaminant in isolation from all others. The real world is much more complex, and standards must be based upon combinations of contaminants as well as our individual susceptibilities to adverse reactions. We are still in the dark when it comes to the necessary understanding of biological response to contaminants. Therefore, we must err on the side of caution and require more restrictive purity standards for all drinking water.

PCBs Revisited

After nearly a generation of corporate delays and governmental "hop-scotch" tactics, the toxic and carcinogenic PCB contamination in one part of our country may be *partially* cleaned up soon. On April 12, 2006, Wisconsin's Governor Jim Doyle announced that an agreement had been achieved with two paper companies, NCR Corp. and Sonoco U.S. Mills, Inc., to start removing PCBs from some "hot spot" Fox River sediments in northeastern Wisconsin beginning in 2007. The two companies will provide approximately $30 million for the dredging and landfill disposal of some of the PCBs in the Fox River. This represents less than 15% of the total estimated PCB contamination now in the Fox River.

The eventual removal of all PCB contamination will require at least a decade of work and many hundreds of millions of dollars more. Meanwhile, the people of northeastern Wisconsin shoreline communities, including those in Green Bay, Oneida, Appleton, and De Pere, continue to receive advisories to avoid consuming fish from the contaminated waters.

However, what about the public consumption and recreational use of the water itself? Shouldn't the public also be concerned about the root sources? How long must we wait for corporate-caused water contamination and recreational waters to be detoxified? So far, more than two generations of men, women, and children, and many more generations of animal life, will have suffered the consequences of exposure to PCB-contaminated water. How many more generations will suffer? We the public, the consumers of corporate products and those who pay the taxes for necessary government services, must speak out. We must demand effective action to protect our water supplies.

Population Density and Contamination

As more land is developed and as population increases, so does contamination. The Environmental Data Resources (EDR) group, located in Milford, Connecticut, released a study in April 2006 that confirmed this rather self-evident observation. Their report summarizes findings made in California, our most populous state, identifying nearly 36,000 contaminated sites of toxic water contamination. These include nearly 21,000 known leaking storage tank sites that are contaminating drinking water aquifers, and over 400 Superfund sites that contain hazardous wastes (mainly from industrial sources) that require remediation. More populated regions were most contaminated. Apparently prompted by these findings, a California legislator has introduced a bill (AB 2228), which if enacted into law would allow property buyers to *purchase* an environmental report that would identify the proximity of toxic sites in nearby areas. Isn't this a wonderful privilege for the people of California! If the bill were to become a law, it would permit a buyer to pay for information that should be required, made available free. Are you curious as to what lobbying groups are involved in this farce? Do you believe that they will allow this bill to see the light of day? Keep posted as the political charade unfolds.

Postscript

The above current events are a snippet of the daily occurrences pertaining to water contamination in our backyards. What happens in another's back yard can also happen in yours. We are all involved whether we want to be or not. We must keep informed about environmental news, particularly as they affect our local community, and take necessary actions to prevent additional contamination. Your health depends, in part, upon your willingness to become informed and involved.

An excellent source of public environmental information is the on-line news-letter *Above-the-Fold* (Reference 8 in the Appendix). This comprehensive publication will open new channels of information to help guide your actions. Subscribe to this free newsletter and read it daily. Also, select your most pressing environmental concern or concerns and get free on-line subscriptions through Google Alerts.

Keep informed, and keep working on behalf of your environment and the improved purity of your drinking water.

8

Pure Water: Your Options

In the U.S., we expect to have the best of everything available. Our drinking water quality is not an exception. All of us, regardless of our economic status or geographical location, expect to have uncontaminated water whenever we open a faucet. Our expectations are still unfulfilled. Certainly compared to "emerging" countries in the world we have much better water quality, but ours is unacceptably contaminated and becoming worse.

The EPA has been around for over thirty-five years, and various "clean something or another" acts have been around for nearly that long. Updated versions of new acts and regulations that flow out of Congress and various agencies *should* be improving water quality and our environment. The facts demonstrate that this has not occurred. The EWG (Appendix, Reference 1) report alone is sufficient evidence that a major problem exists. Combine that with the human contamination described in the U.S. Federal Government's Department of Health and Human Services report (Appendix, Reference 2), and the substantiated and likely health consequences (Appendix, References 3, 4, 5). The EPA and federal government's deficiencies in correcting acknowledged problems are inexcusable. Very recent events prove that our government continues to ignore the contamination and the health problems even as they worsen.

More Government Actions Degrading Water Quality

In 2002, the Bush Administration withheld a report prepared by the Center for Disease Control (CDC) that recommended decreasing the allowable lead concentrations in drinking water. The report showed that the EPA's allowable levels were excessive and that as a result, children and adults had begun to manifest serious brain and central nervous system problems. Political appointees to the

CDC who were employees of paint companies that used lead in their products were able to block the publication of the report.

The Bush Administration also suppressed a report by the U.S. Department of Agriculture that described the potential for the airborne release of disease-causing bacteria from farm waste matter (particularly, from factory farms).

In another example, the Bush EPA withheld its own study from the Senate that stated that unless emissions of mercury, nitrogen oxides, sulfur dioxide, and carbon dioxide were reduced in the U.S., millions of additional Americans would suffer from diseases, including respiratory impairment, heart disease, and stroke.

The Bush Administration has rolled back, stopped, and relaxed the enforcement of earlier environmental regulations that the Clinton Administration had implemented to reduce industrially generated pollution and the health problems that it causes.

A particularly appalling example of the Bush Administration's continual efforts to ditch environmental protection regulation pertains to mercury emissions. Mercury is an established toxin that causes brain damage, neurological impairment, birth abnormalities, and mental and ambulatory degradation in children. Yet, President Bush and his politically-controlled EPA refuse to impose technologically attainable requirements upon coal-burning power plants that would lower mercury emissions by 90%. In February of 2005, the EPA Inspector General reported that EPA agency scientists had been "pressured" to change their scientific findings regarding the damage that mercury inflicts on brain development in young children and fetuses. Once again, the lobbyists and energy industry contributors to the Bush political pocket thumbed their noses at the American people by favoring corporate influence over the protection of people.

Another inexplicable action of the Bush Administration's EPA occurred on October 4, 2005 (Federal Register, Vol. 70, No. 191, TRI-2005-0073), when the EPA proposed revisions for industry reporting procedures for toxic releases into the environment. As of January 16, 2006, the proposed revisions have not been implemented. Many environmental groups have objected. However, implementation is likely within the first half of 2006. Any company operation is currently required to submit a report, referred to as the Toxic Release Inventory (TRI), to the EPA listing any toxic discharge that exceeds 500 pounds annually. The proposed revision changes the allowed limit to 5,000 pounds annually. This ten-fold increase will effectively eliminate public disclosure of thousands of deliberate toxic releases. Why would the EPA do something that consciously prevents the public disclosure of toxic contaminant released into the environment? Many companies will view the relaxed reporting requirements as an open invitation to contaminate more. What corporate lobbying groups, and what members of the Bush Administration, are responsible for this apparent disdain for the public health?

Independent Oversight and Actions Are Needed

Since our government, particularly the current Bush Administration, has repeatedly failed to protect our water quality. We must assume responsibility to do so ourselves until diligent and effective government becomes responsive to the people's needs. This presents a substantial problem. Our primary need is to ensure that the water we consume is free from harmful contaminants. This is not a trivial problem. Our normal tap water is likely to contain numerous contaminants, both regulated and unregulated (and generally unidentified), if it is from a municipal supply. Contaminants contained in water from private wells are generally unknown and seldom analyzed, except where unusual water quality problems have been publicized. Furthermore, the cost to obtain comprehensive chemical analyses of our water is economically problematic and impractical, since the concentrations and types of contaminants are likely to change with time. Pinpointing chemical analyses for any specific contaminant provides an incomplete and distorted picture of reality. In our mobile lifestyles, we are forced to use whatever tap water is available wherever we go, in both our food and our drinks. What, therefore, are the best practical options for maximizing our use of safe water and protecting our health from damaging water contaminants until our government (federal, state, and local) really begins to protect us?

Three options are apparent:

1.   Drink and use what comes out of taps wherever you are, and assume the risk.

2.   Purify all water that enters your mouth and washes your body.

3.   Use purified or natural bottled water that a reliable source has determined to be "free from contamination."

Option 1 seems foolhardy if you are economically able to consider either 2 or 3 However, the latter are not slam-dunk options either, since both technical and practical considerations are involved.

Self-Protection

Options 2 and 3 represent individual actions that we should consider if economically feasible. These options, involving a dependence on pure bottled water or water purified using some type of residential or stand-alone system, will help you avoid some contaminants. The purity of bottled water remains a problem. Contact the bottler and obtain information from them about test results and the frequency of testing. Question the results of any report, especially with respect to the scope of the analyses. Make certain that the 119 EPA regulated chemical compounds are included in the report, and ask about analyses of unregulated substances. Make sure that your choice of a specific brand of bottled water is based upon factual data, not marketing ballyhoo. Contact your local Department of Health or the FDA for information. Likewise, do the same when selecting a water purification system or device for your home or small business. Obtain independent information from your local university, college, library, or another source besides the manufacturer or installer of your purification system. Always take the time and effort to learn from an independent and impartial non-commercial source. Independent testing labs can also be useful and impartial.

Private Wells

Those who depend upon private well water should periodically have their water analyzed for contaminants, both organic and inorganic. Consult your local state university extension department preferably the engineering or chemistry chairperson to obtain advice on the contaminants you should have analyzed in your particular area. Take appropriate action to remove the detected contaminants considered harmful to your health. Obtain additional chemical analyses for harmful unregulated contaminants likely to be found in your community. Once again, consult your local university extension or university, where you are most likely to get impartial and independent advice.

Harmfulness depends upon each person's individual susceptibility. Infants and children are generally more susceptible than adults because of their incomplete biological maturation and smaller body mass. Individual chronic allergies and diseases may also increase one's susceptibility to harm from specific contaminants. The water purity you require may create a special situation demanding greater purification than commercial purification systems or bottled water provides. Consult a medical specialist or an allergy/environmental specialists for recommendations regarding your particular medical needs. Seek independent advice from companies that supply different purification systems to determine the type that best serves your needs. One type does not fit all.

"Water Action Group" (WAG)

Option 1 is the only option that many of us can consider because of economic constraints. If so, several other actions are worth consideration. If you use municipal water, request a Water Quality Report from your municipality or Regional Water District and read it thoroughly. *Write* to your water department administrator and ask him or her what the department is doing to eliminate present contaminants, both regulated and unregulated, that are listed in the report, as well as those not listed but suspected. Even if regulated contaminants are at or below a specified MCL, they are still not healthful. Request information from your water department administrator about the sources of the contaminants, and ask who is responsible for their release into the environment. Individual actions are needed but insufficient. Corrective action of some type is also necessary. A possible approach involves beginning a people's movement, based upon collective local action. The following idea may be useful.

Ask your neighbors to join you and form a neighborhood "Water Action Group" (WAG) to work for improved and safer water quality. Learn as much as you can about the health hazards of the contaminants in water supplies and enlist the assistance of willing medical professionals and others in your community to help inform you. Create action plans with your Water Action Group. Make your plan as detailed as you can. State your purpose and goal. List specific actions that are required and the individual assigned to the action. Establish a timetable. Compile the results and write a summary of your findings. Submit them to your local and state elected politicians and environmental departments, asking for their review, comments, and response to your groups needs.

Ask them for help to improve your water quality. Persevere! Keep requesting responses from the government and politicians in your area. Tell the local media about your group's activities and publicize the WAG and its purpose. Help create *local*, *state*, and *national* organizations! We the people can and must help improve the purity of everyone's drinking water. Whatever else you may do to help assure your water quality and health, form or participate in a WAG. Inspire others to join, and develop people power to induce the politicians and corporations to listen.

Self-Protection Plus

Options 2 and 3 represent individual actions that we should perform if they are economically feasible. These options, involving the dependence on pure bottled water or purified water using some type of residential or stand-alone system, will help you avoid some contaminants. However, dependence upon either one is insufficient alone. It may even lead to a general worsening of water contamination due to the sense of individual immunity from contaminated water. This could generate widespread public indifference and apathy, and the government and corporations will allow (or cause) further contamination without fear of backlash. The result will be the continual degradation of water purity, which will reduce the effectiveness of purification treatments, increase cost of treatments and maintenance requirements, and even limit the availability and purity of bottled water. Therefore, options 2 and 3, although protective in the short term, must not become our only actions. Limiting action to your own self-protection is necessary but inadequate and even selfish. This "every man for himself" mentality is the same type of behavior that the corporate/government "cartel" has engaged in for many years. Join a WAG, or help form one, and work to protect water purity for everyone, including yourself.

Your Action Counts

Rachael Carson's work demonstrates what one individual can accomplish. She withstood verbal attacks and ridicule by chemical industry executives, who attempted to belittle her ecological observations and conclusions. She was right. They were wrong. Her legacy is a hallmark of the environmental movement in our country. Unfortunately, her early death at the age of fifty-seven (1964) prevented her from realizing the benefits of her courageous fight on behalf of our environment, and particularly our water resources.

We must all wake up to the deception and underhanded actions of all governments, and particularly the Bush Administration, to weaken environmental regulations. Their corporate puppet masters are pulling the strings, and they will continue to do so until we, the American people, respond by making environmental protection a high priority. Political slogans are nothing but hollow words for the gullible. Stand up and show them that you are not gullible. Your action counts!

9

The Future Depends on Us: Mission Earth

Excessive Development

We humans tend to flock together, forming high-density populations. This tendency results in ever-expanding cities where every square foot of land is covered with roads, buildings, and homes that eventually consume most of the open natural space. Populations skyrocket, and we unconsciously release the manmade contaminants that have become a part of "normal life" into our immediate environment. Businesses and utilities do the same, but they also often create streams of production wastes that they dispose of by emitting them into our atmosphere or discharging them into surfaces or ground water, or both. As a city spreads, the amount of waste mushrooms, and our ability to prevent or avoid contamination diminishes. Our asphalt streets, concrete walkways, solid rooftops, surfaces of engulfed parkland and lakes, and air become laden with contaminants from many sources. Some of these are a small as cigarette remains that we flick out, oil leakage and exhaust from autos and trucks, liquids and substances that we spill or use to clean surfaces, and herbicides, insecticides, and pesticides that we employ to grow more plush plants and lawns on the little bit of land that remains undeveloped.

Other contaminants include such seemingly obscure and seemingly miniscule materials such as shoe sole treads, rubber dust from tires, and myriad plastics, paper, and food remnants. The human waste that we flush down toilets and sinks, as well as the excrement from pets, all adds to the burden of contamination in our immediate environment. Rains and snows carry contaminants from the air to the streets and rooftops, and they wash the contaminants from surfaces into land or storm drains where they eventually migrate into water sources, which become more concentrated with contaminants for our future consumption. This adds up to a veritable cesspool that inevitably contaminates the air we breathe,

the water we drink, and the body we are. There are no easy or immediate solutions, but we can change our lifestyles, begin to restore sanity to our lives, and eventually ease the burden on future generations, at the very least.

First, we must stop our obsession with developing every square foot of land as our communities and cities enlarge. We must allow vast areas of undeveloped and natural areas to remain so that the contaminants that we do generate become less concentrated and somewhat less damaging.

The Burnham plan in Chicago is an example. A visionary architect and land planner, Daniel H. Burnham, was commissioned in 1907 by a group of businessmen known as The Merchants Club to develop a plan for the city to restore the lakeshore and provide parks for the benefit of the people. In 1909, Chicago's-mayor, Fred A. Busse, formed the Chicago Plan Commission. This freed up tax dollars to develop public parks and open spaces along the shoreline. This visionary action, created by Burnham and fostered by businesses and politicians, helped to create numerous open public parks along the Lake Michigan shorelines and throughout Chicago. Burnham's vision—"A very high purpose will be served if the lake shore be restored to the people and made beautiful for them"—became a reality, and it now represents one of the finest examples of land preservation within a large city. Unfortunately for Chicago and most of the communities that have since spawned around it, the Burnham plan was not the standard, as more land was developed and urban sprawl dominated the landscape. Although the Burnham Plan was conceived of solely for aesthetic reasons, it became, nonetheless, a hallmark for lakeshore preservation and the environmentally derived health benefits. The next time you visit Chicago, walk through the parks along the city's shoreline, including what is now Millennium Park, the lakeshore's gem, and remember that unique period when land preservation was advocated and supported by business people and politicians alike for the benefit of all.

Unfortunately, very few communities have followed Burnham's lead for land preservation. Virtually all U.S. cities and municipalities that have grown from their agricultural roots in the early 1900s into the contaminated municipalities of the twenty-first century have failed to create "green" buffers for aesthetic or environmental reasons. Many small towns and villages have become emulators of their big city brothers and sisters, allowing land development for commercial and residential purposes at the expense of our environmental health. The mighty dollar rules and unlimited development prevails.

Natural Land Preservation

We who live in the U.S. are blessed to be in a country that has substantial land available to accommodate population growth without overcrowding and over-development, which causes contamination and the destruction of open natural land. If we are unable or unwilling to modify our desire to concentrate rather than to disperse as our population increases, we will continue to severely contaminate our land, our water, our air, and ourselves. Whether private landowners and public interests can accomplish this change is problematic. Since the value of all land is determined by supply-and-demand factors, the desirable locations become valuable sources of revenue for private land owners, and the human tendency is to squeeze as much revenue as possible from the development of every square foot of land. This becomes the ultimate driving force in land development and most commercial endeavors.

However, the preservation of land for permanent open space and environmental buffers is essential to ensure environmental quality. Governments can and must play a role by allocating funds to acquire private land for preservation as permanent public land in growth areas. Property and real estate tax revenues (including sales transfer fees and impact fees) should help fund community land trusts for the acquisition and preservation of land. This can be a way to develop buffers of permanent open land. This will require the will of both politicians and taxpayers. Voluntary acquisition and the setting aside of land for natural preservation requires funding from individuals to organizations such as the Nature Conservancy and various community-based land trust organizations. Individual landowner-imposed restrictions and effective zoning limitations can also help create permanent land-use restrictions that will reduce development. The state of Wisconsin's "Stewardship Fund" is an effective program to acquire and preserve open lands using public funds.

All of these options, and others, will help to disperse developments and minimize the concentration of environmental contamination and health impairment.

If possible, we must participate in promoting these actions, but at the very least we must become informed so we can vote for politicians who support our environmental protection needs. Resist being influenced by slick words and slogans that are unsupported by tangible actions, beliefs, and voting records (if available) on environmental issues. Write and ask what they will do and what they have done to help ensure a safe environment. Read the words they have written and spoken and examine their actions. Determine what their political party has advocated in the past. Get informed and vote for non-contaminated water and a clean and safe environment. Our health depends upon the attainment of this goal.

Eliminate the Use of Unnecessary "Stuff" and Energy

In addition to land preservation for environmental buffers and aesthetic purpose, we can help prevent additional water contamination by minimizing our use of unnecessary toxic "stuff." For example, stop using toxic herbicides, pesticides, and fungicides in your garden and around your home. Grow plants that are more naturally resistant to insects and bugs. Tolerate weeds or pull them out rather than spraying with a toxic chemical substance. Eliminate artificial fertilizers and use selected plants and vegetation to generate nitrogen in the soil by natural nitrogen fixation from the air. Mulch rather than fertilize and let recycled plants do the job.

Avoid the use of toxic household cleaners and other substances. Instead of using air fresheners, grow indoor plants to remove carbon dioxide and generate oxygen while creating a pleasant and naturally aromatic environment.

Maintain your autos, trucks, boats, lawn mowers, snow blowers as well as your recreational vehicles to avoid leakage of oil and gasoline as well as other hazardous fluids. Do not dispose of any of these fluids by dumping them onto soil or nearby surface water. Power boats, snowmobiles, and other types of internal combustion powered machines are particularly important to maintain properly, since they may leak toxic fluids directly into the water sources that we use.

Exhaust products from the internal combustion engines that power most forms of transportation contain nitrogen oxides, unburned hydrocarbons, and particulate materials that are toxic. Coal-burning power plants that generate electricity also produce emissions that are toxic and become water contaminants. Nuclear power plants for electrical power are sources of highly toxic and carcinogenic wastes that must be disposed of with great care to prevent contamination. Accidental leakage of radioactive cooling water is particularly harmful and insidious, since it may blend with normal water undetected. All of these water contamination sources must be controlled and limited. Do your part by conserving energy use and reducing fuel consumption.

Eliminate the use of personal care items that contain manmade chemicals, such as antiperspirants, hair sprays, and underarm deodorants, and use products that contain only natural substances. Favor certified organic substances and products whenever possible so that you know the manufacturers are not using herbicides, pesticides, fungicides, artificial fertilizers, or other manmade chemicals in their products.

Examine everything that you use and assure yourself that it is as free from manmade chemical contamination as possible. Be selective in what you buy and how you use what you purchase. Develop a pattern of conscious behavior that questions the content of *everything* you use, and avoid using anything that seems suspicious until you get the facts about what it contains and the possible toxicity it may impart to the environment.

Choose energy-efficient appliances, home designs, and materials. When buying a car, put fuel efficiency at the top of your criteria, not only to reduce our oil dependence and costs but, even more importantly, to minimize environmental damage, global warming, and water, air, and food contamination.

Energy conservation helps reduce the contamination that permeates our water supply. We protect our health as we protect our environment. We will also help stabilize our climate by reducing the emissions of chemical substances that increase global warming.

Multiple Benefits

We can all help minimize the sources of contamination by reducing energy consumption and by relying upon energy-efficient processes and products. Certain *hybrid automobiles* can reduce oil consumption by at least 50%. *Wind and solar power* can eliminate most energy-caused contamination and must be the future direction. *Fuel cells* can generate power and pure water at the same time. All of these and other energy-efficient technologies have the promise of improving environmental and water quality and reducing our dependence on fossil fuels. However, as with all new technologies, there will be unavoidable incubation periods during which research and development costs must be borne. Private sources have often done this in the past, but in this era of greed and instant corporate returns on invested dollars, it is very unlikely that the required new technology will become commercialized without a massive government program supported by our taxes. However, the new horizon will become visible if we commit ourselves to increasing renewable energy sources and reducing our dependence upon fossil fuels.

Could there be a better cause? In addition to all the reasons cited above, you will be contributing to general economic improvement as well. Substitution of non-toxic substances will provide economic advantages and jobs to manufacture them with corporations and individuals who participate. Energy conservation technology will create numerous new jobs that will pay more than those in traditional service sectors . Such a scenario is precisely what occurred following the advent of manned space exploration in the 1960s. New technology generated millions of well paying jobs. This is a win-win scenario for everyone—it benefits our environment, our economy, and our health.

Mission Earth

During the early 1960s, our country was inspired, mobilized, and led by President Kennedy to send man into space and to the moon. As a young engineer working in aerospace, I had the good fortune of being directly involved in this great adventure. The enormous feat of landing astronauts on the moon became reality within ten years of President Kennedy's statement of purpose. Our country demonstrated what a great country, under great leadership, could accomplish. The technological spin-offs and high-paying jobs were enormous and included innumerable computer, medical, and electronic breakthroughs, and associated economic opportunities. The same types of opportunities are attainable today, and for an even more vital, Earth-bound purpose. But first we need the leadership and will of our politicians and the encouragement and support of our people.

Instead of "Mission Space," which President Kennedy fostered, "Mission Earth" must become the new priority of visionary political leaders. They must work for an environmental future that will benefit not only our country, but the whole world. We have the technical capacity to accomplish such a mission within a decade or two if we mobilize our resources, ingenuity, and the spirit and ingenuity of our people that made our country great. If we have the will, we can all but eliminate our dependence on fossil fuels. We can begin to restore our environment to conditions that prevailed seventy-five years ago or more if we are committed to accomplishing this task. Many millions of new employment opportunities are certain, and technological innovations for the benefit of our present and future generations will bloom. We will begin to mobilize our collective spirit and ingenuity for our sake and the sake of future generations. We will have to shift our priorities and budgets to fund the necessary programs and research, just as was done during the space program to land man on the moon. This time our focus must be on earth.

This may seem utopian, but our world can change. We can reduce suffering from the effects of devastating poverty, destructive wars, starving masses, diseases and deaths caused by global contamination, and potentially calamitous climate changes induced by global warming. We can become a world where poverty and starvation are unknown, where all life is in balance and harmony with nature, and where pure water, pure air, and nature's open and non-contaminated lands become all humanity's gifts to one another. This vision is attainable. *The future depends on us.* Get on board "Mission Earth."

Conclusion

Once contaminants become part of our body (from water, air, tactile means, and food), they are transferred into our blood and then to virtually every tissue, bone, and organ in our body. Thousands of miles of blood vessels and capillaries distribute blood, which contains cellular oxygen and other essential components (platelets, white cells, etc.), to every cell in our body. Unnatural and potentially harmful contaminants also make this journey into our cells. Some of the contaminants are very stable, and they remain in our organs, bones, and tissues. Some contaminants are partially or completely eliminated by our body's natural waste-elimination system as impurities in urine, feces, and perspiration. Those retained may pose a greater disease risk over time, as accumulation increases. Transient exposures can also be damaging if the frequency or severity of exposure is high; even brief durations are dangerous in some cases.

The progressive contamination of Earth's water by human actions will inevitably cause the world's water supply to become deadly. At some time in the future, the toxicity and biological dangers that will exist will be so great that Earth's water may become too dangerous to consume, and possibly too saturated with contaminants to purify.

The destiny of the world's water and energy supplies will depend upon the actions of world leaders and an enlightened public's participation. The human inclination to procrastinate or to be apathetic about more remote problems is the biggest enemy of enlightened action. We are inclined to "live in the moment" and to "go with the flow." Consequently, we often fail to recognize impending problems until they become nearly unmanageable. Ignoring the current and impending water contamination problems will be fatal to most, if not all, life on Earth. Before deciding what actions are appropriate, however, it is useful to review some of the facts about water and its contamination. That's what this little book is all about.

Approximately seventy-five years ago, the world's population was less than two billion people, and today the global population is over 6 billion. The world's population will probably exceed 10 billion before 2020. The population of the U.S. will probably grow from about 290 million at present to over 400 million by 2020. Increasing populations, existing and new manmade chemical substances released into the Earth's waters and environment, and a limited supply of water for all Earth's people will cause increasing toxicity and pollution of the world's water supplies. The consequences are likely to be disastrous. Human ingenuity and our passion to preserve life must prevail. The future is in our hands, and we must enable Nature to prevail. The Manmade Toxic Contamination Age must end. A new era of pure water must begin.

APPENDIX

SELECTED REFERENCE SOURCES

1. Environment Working Group. *A National Assessment of Tap Water Quality.* December 20, 2005.

2. Department of Health and Human Services, Centers for Disease Control and Prevention. *Third National Report on Human Exposure to Environmental Chemicals.* July 2005.

3. Division of Cancer Epidemiology and Genetics, National Cancer Institute, et al. *Drinking Water: Nitrate and Health.* June 23, 2005.

4. Boston University School of Public Health and the Environmental Health Initiative. *Environmental and Occupational Causes of Cancer.* University of Massachusetts, Lowell, September 2005.

5. *National Toxicology Program: The Report on Carcinogens*, 11th ed. January 31, 2005.

6. *The Quality of Our Nation's Waters: Nutrients and Pesticides.* U.S. Geological Survey Circular 1225, 1999.

7. Harte, John, et al. *Toxics A to Z: A Guide to Everyday Pollution Hazards.* University of California Press, 1991.

8. "The World Factbook," online at www.cia.gov/cia/publications/factbook/geos/us.html

9. Moss, Brian. *Ecology of Fresh Waters*, 3rd ed. Blackwell Science, Ltd., 1998.

10. Leopold, Aldo. *A Sand County Almanac.* Oxford University Press, 1949.

11. Carson, Rachel. *Silent Spring.* Houghton Mifflin Company, 1994 (introduction by Al Gore).

12. Kennedy, Robert F., Jr. *Crimes against Nature*. Harper Collins Publishers, 2004.

13. U.S. Department of the Interior. *Contamination Potential in Silurian Dolomite. Door County*. Paper 2047, 1978.

14. Environmental Working Group Chemical Archives, online at <u>www.chemicalindustryarchives.org</u>

AN EXAMPLE OF AN ACTUAL DRINKING WATER QUALITY REPORT

TEST RESULTS

Contaminant	Violation Y/N	Level Detected	Unit Measurement	MCLG	MCL	Likely Source of Contamination
Microbiological Contaminants						
1. Total Coliform Bacteria	N	0		0	presence of coliform bacteria in 5% of monthly samples	Naturally present in the environment
2. Fecal coliform and E.coli	N	0		0	a routine sample and repeat sample are total coliform positive, and one is also fecal coliform or E. coli positive	Human and animal fecal waste
3. Turbidity	N	.3	n/a	TT		Soil runoff
Radioactive Contaminants						
4. Beta/photon emitters			mrem/yr	0	4	Decay of natural and man-made deposits
5. Alpha emitters	N	2	pCi/l	0	15	Erosion of natural deposits
6. Combined radium	N	1.8	pCi/l	0	5	Erosion of natural deposits
Inorganic Contaminants						
7. Antimony	N	ND	ppb	6	6	Discharge from petroleum refineries; fire retardants; ceramics; electronics; solder
8. Arsenic	N	ND	ppb	n/a	50	Erosion of natural deposits; runoff from orchards; runoff from glass and electronics production wastes
9. Asbestos			MFL	7	7	Decay of asbestos cement water mains; erosion of natural deposits
10. Barium	N	.049	ppm	2	2	Discharge of drilling wastes; discharge from metal refineries; erosion of natural deposits
11. Beryllium	N	ND	ppb	4	4	Discharge from metal refineries and coal-burning factories; discharge from electrical, aerospace, and defense industries
12. Cadmium	N	ND	ppb	5	5	Corrosion of galvanized pipes; erosion of natural deposits; discharge from metal refineries; runoff from waste batteries and paints
13. Chromium	N	ND	ppb	100	100	Discharge from steel and pulp mills; erosion of natural deposits
14. Copper	N	.99	ppm	1.3	AL=1.3	Corrosion of household plumbing systems, erosion of natural deposits; leaching from wood preservatives
15. Cyanide			ppb	200	200	Discharge from steel/metal factories; discharge from plastic and fertilizer factories
16. Fluoride	N	.57	ppm	4	4	Erosion of natural deposits; water additive which promotes strong teeth; discharge from fertilizer and aluminum factories
17. Lead	N	4.3	ppb	0	AL=15	Corrosion of household plumbing systems, erosion of natural deposits
18. Mercury (inorganic)	N	ND	ppb	2	2	Erosion of natural deposits; discharge from refineries and factories; runoff from landfills; runoff from cropland
19. Nitrate (as Nitrogen)	N	.95	ppm	10	10	Runoff from fertilizer use; leaching from septic tanks, sewage; erosion of natural deposits
20. Nitrate (as Nitrogen)	N	ND	ppm	1	1	Runoff from fertilizer use; leaching from septic tanks, sewage; erosion of natural deposits
21. Selenium	N	ND	ppb	50	50	Discharge from petroleum and metal refineries; erosion of natural deposits; discharge from mines
22. Sodium	N	14.00	ppm	n/a	n/a	Leaching from ore-processing sites; discharge from electronics, glass, and drug factories
23. Thallium			ppb	0.5	2	
Synthetic Organic Contaminants including Pesticides and Herbicides						
24. 2,4-D	N	ND	ppb	70	70	Runoff from herbicide used on row crops
25. 2,4,5-TP (Silvex)	N	ND	ppb	50	50	Residue of banned herbicide
26. Acrylamide				0	TT	Added to water during sewage/waste water treatment
27. Alachlor	N	ND	ppb	0	2	Runoff from herbicide used on row crops
28. Atrazine	N	ND	ppb	3	3	Runoff from herbicide used on row crops
29. Benzo(a)pyrene (PAH)	N	ND	nanograms/l	0	200	Leaching from linings of water storage tanks and distribution lines

Water Quality Report

Contaminant	Violation Y/N	Level Detected	Unit Measurement	MCLG	MCL	Likely Source of Contamination
30. Carbofuran	N	ND	ppb	40	40	Leaching of soil fumigant used on rice and alfalfa
31. Chlordane	N	ND	ppb	0	2	Residue of banned termiticide
32. Dalapon	N	ND	ppb	200	200	Runoff from herbicide used on rights of way
33. Di(2-ethylhexyl) adipate	N	ND	ppb	400	400	Discharge from chemical factories
34. Di(2-ethylhexyl) prithalate	N	ND	ppb	0	6	Discharge from rubber and chemical factories
35. Dibromochloropropane	N	ND	nanograms/l	0	200	Runoff/leaching from soil fumigant used on soybeans, cotton, pineapples, and orchards
36. Dinoseb	N	ND	ppb	7	7	Runoff from herbicide used on soybeans & vegetables
37. Diquat	N	ND	ppb	20	20	Runoff from herbicide use
38. Dioxin (2,3,7,8-TCDD)			picograms/l	0	30	Emissions from waste incineration and other combustion; discharge from chemical factories
39. Endothall	N	ND	ppb	100	100	Runoff from herbicide use
40. Endrin	N	ND	ppb	2	2	Residue of banned insecticide
41. Epichlorohydrin				0	TT	Discharge from industrial chemical factories; an impurity of some water treatment chemicals
42. Ethylene dibromide			nanograms/l	0	50	Discharge from petroleum refineries
43. Glyphosate	N	ND	ppb	700	700	Runoff from herbicide use
44. Heptachlor	N	ND	nanograms/l	0	400	Residue of banned termiticide
45. Heptachlor epoxide	N	ND	nanograms/l	0	200	Breakdown of heptachlor
46. Hexachlorobenzene	N	ND	ppb	0	1	Discharge from metal refineries and agricultural chemical factories
47. Hexachlorocyclo-pentadiene	N	ND	ppb	50	50	Discharge from chemical factories
48. Lindane	N	ND	nanograms/l	200	200	Runoff/leaching from insecticide used on cattle, lumber, gardens
49. Methoxychlor	N	ND	ppb	40	40	Runoff/leaching from insecticide used on fruits, vegetables, alfalfa, livestock
50. Oxamyl (Vydate)	N	ND	ppb	200	200	Runoff/leaching from insecticide used on apples, potatoes and tomatoes
51. PCBs (Polychlorinated biphenyls)	N	ND	nanograms/l	0	500	Runoff from landfills; discharge of waste chemicals
52. Pentachlorophenol	N	ND	ppb	0	1	Discharge from wood preserving factories
53. Picloram	N	ND	ppb	500	500	Herbicide runoff
54. Simazine	N	ND	ppb	4	4	Herbicide runoff
55. Toxaphene	N	ND	ppb	0	3	Runoff/leaching from insecticide used on cotton and cattle

Volatile Organic Contaminants

Contaminant	Violation Y/N	Level Detected	Unit Measurement	MCLG	MCL	Likely Source of Contamination
56. Benzene	N	ND	ppb	0	5	Discharge from factories; leaching from gas storage tanks and landfills
57. Carbon tetrachloride	N	ND	ppb	0	5	Discharge from chemical plants and other industrial activities
58. Chlorobenzene	N	ND	ppb	100	100	Discharge from chemical and agricultural chemical factories
59. o-Dichlorobenzene	N	ND	ppb	600	600	Discharge from industrial chemical factories
60. p-Dichlorobenzene	N	ND	ppb	75	75	Discharge from industrial chemical factories
61. 1,2 - Dichloroethane	N	ND	ppb	0	5	Discharge from industrial chemical factories
62. 1,1 - Dichloroethylene	N	ND	ppb	7	7	Discharge from industrial chemical factories
63. cis-1,2-ichloroethylene	N	ND	ppb	70	70	Discharge from industrial chemical factories
64. trans - 1,2-Dichloroethylene	N	ND	ppb	100	100	Discharge from industrial chemical factories
65. Dichloromethane	N	ND	ppb	0	5	Discharge from pharmaceutical and chemical factories
66. 1,2-Dichloropropane	N	ND	ppb	0	5	Discharge from industrial chemical factories
67. Ethylbenzene	N	ND	ppb	700	700	Discharge from petroleum refineries
68. Styrene	N	ND	ppb	100	100	Discharge from rubber and plastic factories; leaching from landfills
69. Tetrachloroethylene	N	ND	ppb	0	5	Leaching from PVC pipes; discharge from factories and dry cleaners
70. 1,2,4 - Trichlorobenzene	N	ND	ppb	70	70	Discharge from textile-finishing factories
71. 1,1,1 - Trichloroethane	N	ND	ppb	200	200	Discharge from metal degreasing sites and other factories
72. 1,1,2 - Trichloroethane	N	ND	ppb	3	5	Discharge from industrial chemical factories
73. Trichloroethylene	N	ND	ppb	0	5	Discharge from metal degreasing sites and other factories
74. TTHM (Total trihalomethanes)	N	3.9	ppb	0	100	By-product of drinking water chlorination
75. Toluene	N	ND	ppm	1	1	Discharge from petroleum factories
76. Vinyl Chloride	N	ND	ppb	0	2	Leaching from PVC piping; discharge from plastics factories
77. Xylenes	N	ND	ppm	10	10	Discharge from petroleum factories; discharge from chemical factories

Unregulated Contaminants

Contaminant	Violation Y/N	Level Detected	Unit Measurement	MCLG	MCL	Likely Source of Contamination
78. Bromodichloromethane	N	1.30	ppb	n/a	n/a	n/a
79. Chloroform	N	1.90	ppb	n/a	n/a	n/a
80. Dibromochloromethane	N	.69	ppb	n/a	n/a	n/a

Water Quality Report

Sister Bay Utilities routinely monitors for constituents in your drinking water according to Federal and State laws. This table shows the results of our monitoring starting January 1st, 1999. All drinking water, including bottled drinking water, may be reasonably expected to contain at least small amounts of some constituents. It's important to remember that the presence of these constituents does not necessarily pose a health risk.

In this table you will find many terms and abbreviations you might not be familiar with. To help you better understand these terms we've provided the following definitions:

Non-Detects (ND) - laboratory analysis indicates that the constituents is not present.

Parts per million (ppm) or Milligrams per liter (mg/l) - one part per million corresponds to one minute in two years or a single penny in $10,000.

Parts per billion (ppb) or Micrograms per liter - one part per billion corresponds to one minute in 2,000 years, or a single penny in $10,000,000.

Parts per trillion (ppt) or Nanograms per liter (nanograms/l) - one part per trillion corresponds to one minute in 2,000,000 years, or a single penny in $10,000,000,000.

Parts per quadrillion (ppq) or Picograms per liter (picograms/l) - one part per quadrillion corresponds to one minute in 2,000,000,000 years or one penny in $10,000,000,000,000.

Picocuries per liter (pCi/L) - picocuries per liter is a measure of the radioactivity in water.

Millirems per year (mrem/yr) - measure of radiation absorbed by the body.

Million Fibers per Liter (MFL) - million fibers per liter is a measure of the presence of asbestos fibers that are longer than 10 micrometers.

Nephelometric Turbidity Unit (NTU) - nephelometric turbidity unit is a measure of the clarity of water. Turbidity in excess of 5 NTU is just noticeable to the average person.

Action Level - the concentration of a contaminant which, if exceeded, triggers treatment or other requirements which a water system must follow.

Treatment Technique (TT) - (mandatory language) A treatment technique is a required process intended to reduce the level of a contaminant in drinking water.

Maximum Contaminant Level - (mandatory language) The "Maximum Allowed" (MCL) is the highest level of a contaminant that is allowed in drinking water. MCLs are set at close to the MCLGs as feasible using the best available treatment technology.

Maximum Contaminant Level Goal - (mandatory language) The "Goal" (MCLG) is the level of a contaminant in drinking water below which there is no known or expected risk to health. MCLGs allow for a margin of safety.

Water Quality Report

Author's Statement of Purpose

✦

By Zalman P. Saperstein

Every day, we turn a water tap and water comes out for us to use. This is almost an automatic action. We believe that the water is always pure and safe.

Is it?

How often do we read about contaminated water that is the cause of health problems? Many of us have experienced water boil orders from local officials because of unexpected bacterial pollution. Various types of chemical pollutants have also been responsible for the temporary or permanent termination of many water supplies. All of these experiences may have prompted you to ask what's behind these drinking water alerts. You may have wondered whether the water your family consumes is safe.

This book explores the question of water safety and presents an historical perspective of why we should be concerned. Corporations have often caused contamination of our drinking water supplies, and governments have often failed to protect the public from such dangers. Exposure to toxins in our drinking water may cause serious health consequences. Government agencies do not protect us from water contamination, as we expect. We as individuals bear a significant responsibility for our water purity.

Ensuring the quality of our drinking water is a responsibility that we all share, and we can all do something about it. Our health and life quality depends upon informed action. This book encourages you to become informed and involved about drinking water issues. This book will give you a useful start and guide you to purer water.

About the Author,
Zalman P. Saperstein

From his childhood in the 1930s until the early 1960s, Zalman lived in Los Angeles, where he experienced the gradual decline of the environment from smog and water contamination. An early essay he wrote while in elementary school described some of the water issues confronting California, as seen by a child's eye. After receiving a BS in engineering at UCLA in 1954, he began graduate studies. Following military service during the Korean War, he received an MS in engineering as part of a multidisciplinary program at UCLA in 1962. He also completed additional graduate work in biotechnology, material sciences, and applied mathematics.

From the 1950s through the 1990s, his professional life centered on research and development work involving petrochemical materials, aerospace rocketry (including manned space flight), metallurgy, heat transfer, and manufacturing processes. Assignments included tenures at C.F. Braun Engineering Company, Douglas Aircraft, and Illinois Institute of Technology Research Institute. From 1978 until 1995, he was director of research and development and vice president of technical services with Modine Manufacturing Company, headquartered in Racine, Wisconsin.

Zalman authored many technical reports, including NASA-sponsored research and peer-reviewed technical journal papers, and contributed to several technical books. He is the inventor or co-inventor of approximately thirty U.S. patents. He is an elected Fellow of the American Society for Metals and an elected member of the Society of Sigma-Xi, an honorary scientific research society.

Contact: zeepsap@gmail.com

978-0-595-39518-7
0-595-39518-X

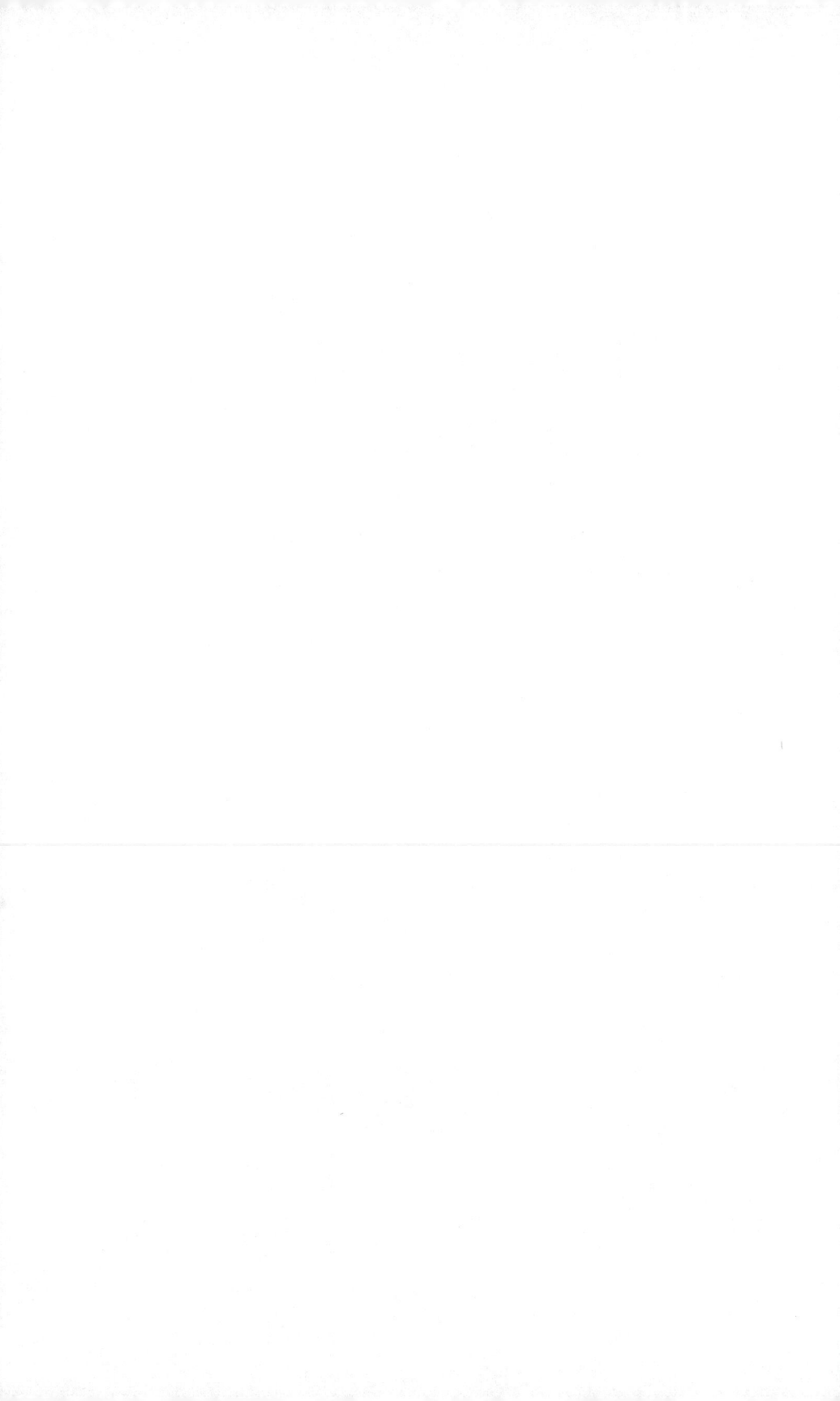

www.ingramcontent.com/pod-product-compliance
Lightning Source LLC
Chambersburg PA
CBHW030346290526
45785CB00004B/1616